Glaciers - Recent Research, Importance to Humanity and the Effects of Climate Change

Edited by Stuart Arthur Harris

Published in London, United Kingdom

Glaciers - Recent Research, Importance to Humanity and the Effects of Climate Change
http://dx.doi.org/10.5772/intechopen.1001624
Edited by Stuart Arthur Harris

Contributors
Anna Orłowska, Binaya Pasakhala, Elliot Moravec, Finu Shrestha, Georges Djoumna, Jay Johnson, Kristina Slawny, Mary R. Albert, Patricio Andino Guarderas, Rodrigo Espinosa Barrera, Sebastian H. Mernild, Sudan Bikash Maharjan, Tanner Kuhl, Tenzing Choyal Sherpa

First published in London, United Kingdom, 2024 by IntechOpen
IntechOpen is the global imprint of INTECHOPEN LIMITED, registered in England and Wales, registration number: 11086078, 167-169 Great Portland Street, London, W1W 5PF, United Kingdom

British Library Cataloguing-in-Publication Data
A catalogue record for this book is available from the British Library

Additional hard and PDF copies can be obtained from orders@intechopen.com

Glaciers - Recent Research, Importance to Humanity and the Effects of Climate Change
Edited by Stuart Arthur Harris
p. cm.
Print ISBN 978-0-85466-794-9
Online ISBN 978-0-85466-793-2
eBook (PDF) ISBN 978-0-85466-795-6

For EU product safety concerns:
IN TECH d.o.o., Prolaz Marije Krucifikse Kozulić 3, 51000 Rijeka, Croatia,
info@intechopen.com or visit our website at intechopen.com.

Meet the editor

Stuart Arthur Harris earned a BSc (Honours), MSc, and Ph.D. in Geology, as well as a DSc in Geography from Queen Mary University, London. He advised the Chief Engineers Branch, British Troops Egypt, and the Arab Legion Engineers in Jordan, solving problems in geology, water supply, and engineering. Subsequently, he was a soil surveyor for Hunting Technical Services, before becoming a soil surveyor for the Government of Guyana. He taught Geography at the University of Chicago, Wilfred Laurier University, and the University of Kansas before joining the University of Calgary in 1969. The Russian Geographical Society awarded him the Nikolai Mihailovich Prjevalsky Medal for his work on Alpine Permafrost in 1996. He has also organized three international field trips in the Canadian Rocky Mountains for international conferences.

Contents

Preface

Glaciers are a well-studied and obvious feature of many of the landscapes on the land surface of the Earth. They develop under cold climatic conditions where there is sufficient precipitation, for example, in polar areas such as Antarctica or in humid mountainous regions such as the Alps or Rocky Mountains. With less precipitation in cold climates, these are replaced by permafrost consisting of ground ice that remains frozen for more than two years. These ice deposits can tell us a lot about past climatic conditions, and when the ice melts, they supply significant quantities of water to the rivers downstream, which is particularly important in semiarid regions.

Climate change has been occurring throughout the history of the Earth [1]. There have been long periods of weather when the mean air and sea temperature has been up to 14°C above that of today, alternating with colder periods lasting up to 60 Ma. During the cold periods, glaciers advance and retreat under the influence of the Milankovitch cycles [2]. We have just finished an interglacial period lasting about 10 ka during which many glaciers have been retreating while the weather has been relatively docile. This changed in 2020, [3] and the weather pattern has become very violent as the cold air mass over Siberia has started to cool due to the beginning of a new 23 ka Precession cycle while the source areas for the subtropical air masses continue to warm. This has resulted in enlargement of the Rossby waves so that they have become the size of Canada and the United States combined. The clash between the colder and hotter air masses produces more violent storms, tornadoes, hurricanes, and the newly recognized atmospheric rivers. Heavy precipitation occurs along the junction of the two air masses, but droughts plague other areas.

This book starts with a chapter entitled "Influence of Atmospheric Rivers on Glaciers", which discusses a subject that has become very important during the last five years. The next section deals with methodology and interpretation of data. The first chapter in this section is "Drilling Ice and Subglacial Rock Cores for Scientific Discovery in a Changing Climate", followed by "Perspective Chapter: What Sort of Ice Dynamics Are Crevasse Fill Ridges Connected with? – Research Overview". The following section includes the chapter, "Spatial Distribution and Shrinkage of Glaciers since the 1990s in the Transboundary Kailash Sacred Landscape", showing an example of the changes in tropical glaciers occurring at the end of an Interglacial event. The final section consists of a chapter on the role of the changes brought on by the Precession cycle entitled "Climate Change in Tropical Glacier-Fed Rivers, Contrasting Global Responses and Future Implications on Stream Functional

Diversity". These chapters provide readers with examples of our state of knowledge as we enter into the beginning of the next glacial event caused by the change in the Precession cycle.

Stuart A. Harris
Department of Geography,
University of Calgary,
Canada

References

[1] Harris SA. Evolution of the climate of the Earth. Journal of Civil Engineering Research and Technology. 2023a;**5**(2):1-15

[2] Milankovitch M. Théorie Mathématique des Phénomènes Produites par la Radiation Solaire. Paris: Gauthier-Villars; 1920. [In French]

[3] Harris SA. Comparison of recently proposed causes of Climate Change. Atmosphere. 2023b;**14**:1244-1260

Section 1

Climate Change and Glaciers

Chapter 1

Influence of Atmospheric Rivers on Glaciers

Georges Djoumna and Sebastian H. Mernild

Abstract

Atmospheric rivers (ARs) are long, narrow, and transient corridors of robust horizontal water vapor transport commonly associated with a low-level jet stream ahead of the cold front of an extratropical cyclone. These weather features are essential for Earth's hydrological cycle, transporting water vapor poleward, delivering precipitation for local climates, and having societal repercussions, such as intense storms and flood risk. The polar regions have experienced increasing AR activity in recent years. ARs usually transport substantial amounts of moisture and heat poleward that can potentially affect glaciers and sea ice. Many studies have demonstrated that ARs cause surface melting of glaciers in Antarctica and Greenland. Predicting and understanding the characteristics of ARs under global warming is a challenging task because there is not a consensus among scientists on a quantitative definition of ARs and the tracking methods. Understanding how ARs affect the surface mass balance of glaciers is crucial to increase our knowledge of how a warming atmosphere associated with warm ocean water will impact glaciated areas. In this work, we review recent advances in AR, including the methods used to identify them, their impacts on glaciers, their relationship with large-scale ocean-atmosphere dynamics, and variabilities under future climate.

Keywords: atmospheric rivers, transport of moisture, warm air intrusion, intense precipitation, extratropical cyclones, climate change

1. Introduction

Over the last decade, interest in atmospheric river (AR) science has significantly increased because ARs are increasingly recognized globally as an essential weather phenomenon associated with extreme precipitation [1–5]. ARs potentially affect the global water vapor distribution and the spatiotemporal structures of the energy and water cycles of the planet [3]. Recent studies suggested that AR significantly impacts the energy and water budgets of the cryosphere, mountain glaciers [6], and polar regions [7–14]. ARs can produce significant snow accumulation over the ice sheet [7, 15–17], melt events with consequences for ice shelf stability [18–21], or calvin events [22]. The American Meteorological Society Glossary defines an AR as "a long, narrow, and transient corridor of strong horizontal water vapor transport typically associated with a low-level jet stream ahead of the cold front of an extratropical cyclone" [1]. ARs account for over 90% of the polar water vapor transport in the midlatitude and, therefore, have important implications for extreme precipitation when they make landfall, affect extreme precipitation events, regional water supply, as well as flooding hazards along western edges of continents (North America, South America, and Western Europe [23]) due to the interaction with the topography. This definition provides a qualitative description of ARs and opens the door to different quantitative definitions that can result in differences in AR climatologies. Understanding ARs in a warming climate is challenging because of the numerous algorithms that have been developed to identify and track the characteristics of ARs. Moreover, the AR community has not agreed on identification methodologies.

Here, we review the impact of AR on sea ice, mountain glaciers, and glaciers in the polar regions. We report on recent advances in AR, including the identification methods, their main climatological characteristics, impacts on glaciers, relationship with large-scale ocean-atmosphere dynamics, and some variabilities under future climate. For a review of the main characteristics of ARs that are responsible for the transport of large amounts of water across the midlatitudes toward higher latitudes, readers are referee to [24]. The climate change impacts of ARs, including physical processes such as the moistening of the atmosphere due to warming, shifting extratropical storm tracks, and the impacts of ARs on the hydrological cycle and hydrologic extremes and crucial AR research directions, extending from the urgency for higher resolution modeling, better observations (especially of regions globally where they are lacking) are reviewed in Payne et al. [4].

Section 2 gives an overview of ARs, their characteristics, their formations, and implications of ARs on sea ice and glaciers in higher latitudes regions and their roles in extreme events and in providing beneficial water supply. The methods used to identify AR are presented in Section 3. Section 4 presents observations and modeling of AR with a focus on high latitudes, while the effect of ARs on sea ice, glaciers, and ice sheets is shown in Section 5. Section 6 addresses AR under a warming climate, and in Section 7, we conclude with suggestions for future research.

2. Overview of atmospheric river and implications

ARs are large-scale weather features that transport significant amounts of moisture that are crucial for large-scale and local hydrological climate across the globe [25]. The impact of ARs on the western coasts of continents in midlatitude locations has been widely studied. In the polar region, emerging research has found that the ARs' related moisture transport can interact with land and sea ice and lead to precipitation events that affect the local cryosphere [7–13, 15–17]. ARs traveling to Antarctica or Greenland from the midlatitudes are usually accompanied by warm air intrusion that can likely affect the sea ice and the stability of ice sheets. A study by Dare et al. [26] displayed a relationship between ARs and the precipitation associated with warm conveyor belt ascent. They argued that precipitation and water vapor transport increase with enhanced moisture in the feeder airstream. ARs can lead to snowfall anomaly, surface melt over glacier, and significantly influence the surface energy balance.

The effect of climate change on ARs or their characteristics under a warming climate is poorly understood. A study by Corringham et al. [27] suggested that ARs are responsible for the majority of the economic losses related to flooding in the Western

United States and are anticipated to intensify in a warming climate. They argued that flood damages caused by extreme AR events may triple from 1 billion to over 3 billion a year toward the end of the century if mitigation actions are not implemented to diminish anthropogenic greenhouse effects. Although the impacts of ARs are increasingly recognized globally, at the global scale, essential questions remain unanswered, such as basic observations with ARs core, the development, and evolution of an AR from its initial setup to dissipation, how an AR interacts with large-scale dynamics, and quantifying the amount of surface melt ARs can trigger over the Arctic and Antarctic [4]. Understanding ARs responses under a warmer climate from a global perspective is of great interest and imperative for predicting weather systems, hydrological extremes, and the preparation for potential associated hazards [28, 29]. An increased trend in AR frequency is demonstrated in climate projection models [28, 30], particularly ARs that make landfall along the California Coast [4, 31]. Moreover, under future warming, it is expected that ARs will become longer, wider, and wetter [32]. For example, a recent study by Michaelis et al. [33] suggested climate change enhanced the precipitation associated with the February 2017 AR over Northern California. However, the physical drivers of AR genesis, development, and dissipation are still poorly understood.

An extratropical cyclone propagates originating from relatively warmer latitudes and propagating across the Southern Ocean is usually characterized by a continuous and strong moisture convergence in the cyclone's warm conveyor belt circulation or along the trailing cold front that feeds the ARs [34–36]. The polar regions have experienced increasing AR activity in recent years [8, 36, 37]. Recently, studies have suggested that sea ice variation in Antarctica and the Arctic can be affected by the poleward transport of moisture and heat from midlatitudes [9, 10, 12, 13, 38]. Recent studies have shown that ARs contribute to the West Antarctic surface melting [11, 21, 39]. Over the West Antarctica Ice Shelf (WAIS), the blocking high and low pressure located along the coast of West Antarctica helps to dive the warm, and moist air to the WAIS. Previous studies have shown that both blocking high- and low-pressure systems are influenced by the El Niño-Southern Oscillation and the Southern Annular Mode and other modes of natural variability that affect the Antarctic continent [25, 40]. Moreover, AR precipitation promotes the rapid increase in surface height over West Antarctica during the 2019 austral winter [15] and abnormal snow accumulation in East Antarctica [7].

In August 2012 and July 2020, Li et al. [10] show that ARs associated with large cyclones triggered rapid sea ice melt through modulating turbulent heat fluxes and winds. They also find a significant negative correlation between atmospheric moisture content and the rate of changes in sea ice concentration over almost the entirety of the Arctic Ocean. The associated warmer air temperature induced by landfalling ARs in the Northeast Greenland ice stream triggers meltwater ponds and rivers that can modify the landscape of the ice stream [41]. The meltwater ponds and rivers on the ice stream absorb more sunlight than the surrounding glacier, and since Northeast Greenland is known to be an area of fast-flowing ice stream draining a large portion of the Greenland Ice Sheet into the ocean, increasing ARs may accelerate the outlet glaciers of this region [41].

ARs also affect the surface hydrological processes of many mountainous regions on the planet. They trigger avalanches and mudslides during winter when they land in mountainous areas [42, 43]. In the Western United States, Hatchett et al. [43] linked the highest percentage of casualties during avalanches to AR conditions. ARs regularly stroke British Columbia (Canada) in the fall and winter, which causes avalanches and heavy precipitation that significantly affect transportation systems [44]. Several studies have demonstrated that ARs triggered extreme precipitation events in mountainous regions with a variety of flooding hazards and ground failures including landslides, and riverbank erosion, for example, [30, 45–50], with a significant impact on the population living in those areas. In the central Hindu-Kush, Karakoram, and Himalaya [46] reported that ARs contribute a significant fraction of the non-monsoon (October–May) precipitation.

2.1 Atmospheric rivers and warm air intrusion

Moist air reaches the Antarctic continent through a limited number of atmospheric processes. The primary one is the advection of warm, marine air. Warm, moist air is a key player in understanding the surface mass balance of the Antarctic [7, 51–54]. Warm air intrusion (WAI) events are more frequent in austral winter in Antarctica than in other seasons. WAI events are crucial in cloud formation and precipitation events on the ASE [55]. Although WAI is frequent in summer, they can trigger surface melting and potentially affect ice sheet stability. ARs and WAI have been documented in the Amundsen Sea Embayment region [14, 40]. Refs. [56, 57] reported heat anomalies from the Atlantic Ocean reached the Greenland Ice Sheet and caused surface melt.

Studying AR and WAI events in the polar region is relevant because surface melted snow or ice can increase ice loss through runoff [58] and also by modifying ice flow dynamics and thermomechanical properties [59, 60]. Moreover, meltwater plays a prominent role in ice shelf hydrofracturing and ice-cliff collapse at deep grounding lines [61]. AR and WAI events are believed to be influenced by regional and large-scale atmospheric variability, including the Southern Annular Mode (SAM) and the El Niño-Southern Oscillation (ENSO), during the summer melt season [62, 63], and it is essential to understand that connection if any.

Figure 1a-d present water vapor imagery obtained from the UW-AMRC repository (ftp://amrc.ssec.wisc.edu/archive/2013/and/2014) during austral summer AR events from 18 to 25 February 2013 (FWAI2013), a fall AR event from 7 to 13 March 2013 (MWAI2013), and early austral winter in the Amundsen Sea Embayment, West Antarctica [14].

3. Theory

ARs are characterized by intense moisture transport. AR intensity is determined by the column-integrated water-vapor transport (IVT),

$$\mathrm{IVT} = \sqrt{\left(\frac{1}{g}\int_{p_t}^{p_s} qudp\right)^2 + \left(\frac{1}{g}\int_{p_t}^{p_s} qvdp\right)^2}, \tag{1}$$

where the gravitational acceleration g is $\mathrm{m\,s^{-2}}$, the specific humidity q is in $\mathrm{kg\,kg^{-1}}$, p is the atmospheric pressure (hPa), p_s surface pressure, and p_t an upper-atmospheric reference pressure (typically set between 500 and 200 hPa, and between 1000 and 300 hPa in the polar regions), and u and v represent the zonal and meridional component of the wind. For the high-latitude regions, vertically integrated water vapor (IWV) fields.

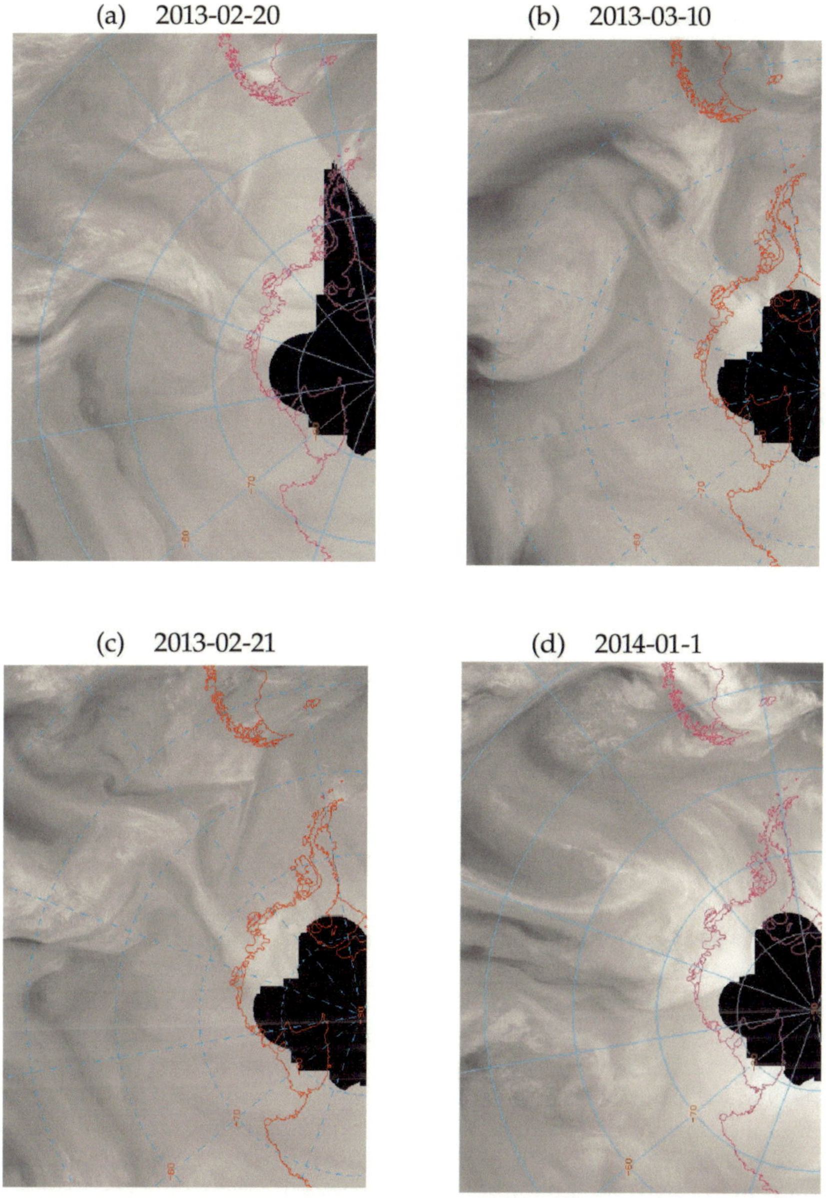

Figure 1.
Composite water vapor imagery data at 5 km spatial resolution on (a) 0000 UTC 20 February 2013, (b) 0000 UTC 10 March 2013, (c) 0012 UTC 21 February 2013, and (d) 0003 UTC 01 June 2014. The data are from the University of Wisconsin Antarctic Meteorological Research Center (UW-AMRC) repository.

$$\mathrm{IWV} = \frac{1}{g}\int_{p_t}^{p_s} q\,dp, \tag{2}$$

are combined with the IVT. Changes in IVT can be partitioned into the change in atmospheric moisture contain and the change in atmospheric motion (through wind vector). The Clausius-Clapeyron equation, which states an exponential-like increase of the water vapor content of saturated air (q^{sat}) with temperature (T), can be employed to investigate the change in the thermodynamic component of IVT. The Clausius-Clapeyron relationship is defined as

$$\frac{dq^{\text{sat}}}{dt} = \alpha(T)q^{\text{sat}}, \tag{3}$$

where $\alpha(T)$ denotes the Clausius-Clapeyron scaling factor, described as

$$\alpha(T) = \frac{L_v}{R_v T^2} \tag{4}$$

where latent heat of fusion L_v and R_w is the gas constant of water vapor [4]. From the Clausius-Clapeyron equation, one can deduce that the fraction of change in the humidity per degree of surface warming is related to the magnitude of warming at a particular height relative to the surface.

$$\frac{1}{\Delta T_s}\frac{\Delta q^{\text{sat}}}{q^{\text{sat}}} = \alpha(T)\frac{\Delta T}{\Delta T_S} \tag{5}$$

The response of IVT at the surface is well understood, while above the surface, $\alpha(T)$ is not constant. An increase in the specific humidity enhances a release of latent heat when saturated air ascends. The lapse rate then decreases with warming and, therefore ΔT increases with height [4].

The response of atmospheric circulation to warming is much less certain [64] and the dynamical responses of IVT that are crucial to the impact of ARs on land are usually their location of landfall and intensity.

3.1 Atmospheric river identification and tracking

Advances in research on ARs have demonstrated their significance for the global water cycle, revealing uncertainties in their mapping, tracking, categorizing, and forecasting landfalling events that make them one of the challenging branches of atmospheric sciences. Numerous methods to identify and track ARs have been developed, usually to address specific research questions. Some criteria used include geometry, threshold values of critical variables, and time dependence, to name a few. These different methods produce differences in AR climatologies and, therefore, differences in the impacts attributable to ARs. More than 20 AR detection methods were assessed in the atmospheric river tracking method intercomparison project [25, 65, 66].

The majority of AR tracking methods often choose a thresholding and are based on the analysis of vertically integrated water vapor transport (IVT) or vertically integrated water vapor (IWV) [25, 66]. The magnitude of the thresholding variable can be either absolute (IVT $\geq$ 250 kg m^{-1} s^{-1}) or relative (IVT $\geq$ 85th percentile of local IVT). Probabilistic IVT forecasts have also been used to determine AR location and intensity [67]. Recent advances in AR tracking methods include machine learning techniques that do not require threshold values have also been employed [68], and a near-global AR detection algorithm that incorporates 3-D wind information from satellite observations [69].

The heat and moisture transport associated with ARs can significantly affect the cryosphere in the polar regions. Indeed, even a slight increase in surface temperature above the freezing point can trigger surface melting over the ice. The impacts of ARs on the Antarctica ice sheet include surface melt in West Antarctica [14, 21], foehn

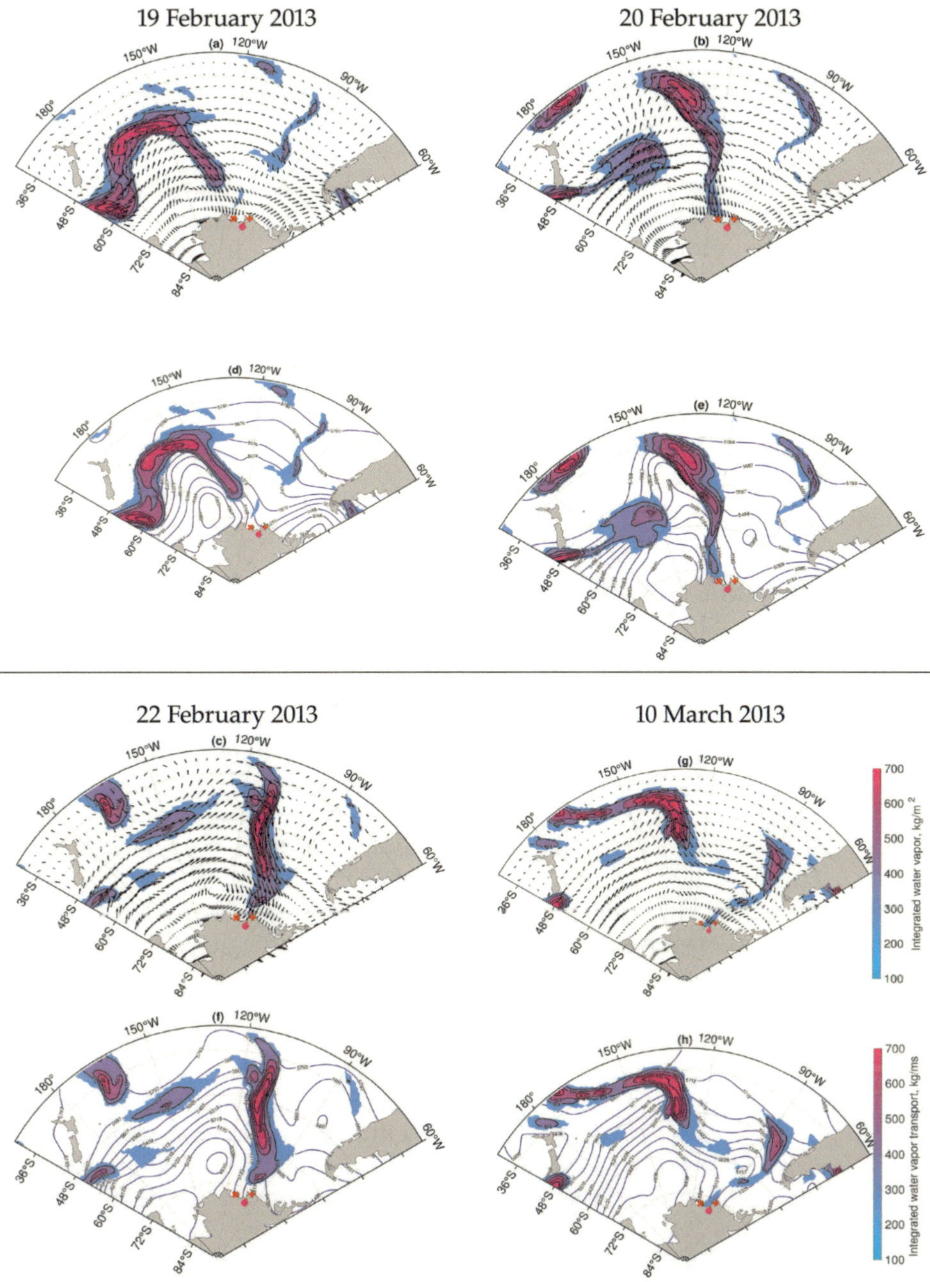

Figure 2.
Panels (a–c) and (g) show the integrated water vapor (IWV, kg m^{-2}*) and the 800 hPa wind vectors obtained from ERA-Interim during two atmospheric rivers that stroke the Amundsen Sea Embayment during 19–22 February 2013 and 10 March 2013. The vertically integrated vapor transport (IVT, kg* m^{-1} s^{-1}*) fields and 500 hPa geopotential heights (m, blue contour lines) are shown in panels (d–f) and (h). The figures are from [14].*

events near the northern tip of the Antarctic Peninsula [39], and snowfall accumulation in East Antarctica [7]. Over the Greenland Ice Sheet, the combined effect of downwelling long-wave radiation and warm air intrusion can negatively impact the ice mass [8, 56, 57].

Djoumna and Holland [14] employed the algorithm developed by Wille et al. [21] for Antarctica to analyze AR events in the Amundsen Sea Embayment (ASE). They combined an IWV-based and IVT-base algorithm between the latitude band of 35°N and 80°S, to locate regions where IVT exceeds IVT's 98th percentile. The IVT threshold of 250 kg m^{-1} s^{-1} was employed to identify AR lateral boundaries as recommended in the literature [7, 39, 70].

Ralph et al. [70] developed a scale to categorize AR events using IVT values that vary from weak AR (IVT$\geq$ 250-500 kg m^{-1} s^{-1}) to extreme AR (IVT$\geq$ 1250 kg m^{-1} s^{-1}). Based on a scale that categorizes AR events based on the maximum instantaneous IVT associated with a period of AR conditions and the duration of those conditions at a point developed by Ralph et al. [70] for the northeastern Pacific Ocean and the Western United States, the MWAI2013 and FWAI2013 AR events in the Amundsen Sea Embayment by Djoumna and Holland [14] were moderate (IVT$\geq$ 500-750 kg m^{-1} s^{-1}) in **Figure 2**. They were weaker than the 25–30 May 2016 event reported by Wille et al. [71].

4. Observations and modeling of AR in the high-latitude regions

Many studies have shown that ARs cause extreme rain events and flooding hazards and can supply water in midlatitude regions. However, the use of numerical weather prediction (NWP) systems to accurately forecast ARs is still under development. Most of the research on the topics has been conducted only during the last two decades. Gaps in the global observing system, for example, across remote ocean areas, and over glaciated areas, hindered efforts to improve forecasts of landfalling ARs. Observation campaigns across AR over glaciers and sea ice are nearly nonexistent [72] used radiosonde measurements collected in austral summer of 2018–2019 (November 2018 to February 2019) at the Dronning Maud Land coast (East Antarctica) to detect AR signatures in the vertical profile of wind speed and direction, air temperature, relative humidity, and specific humidity. Most observation campaigns targeting AR have been conducted in midlatitudes and subtropical regions. Several observation campaigns using the drop-sondes have been conducted in the midlatitudes and subtropical regions; these observations include the vertical profile of water vapor, wind, and pressure and were obtained from 304 aircraft drop-sondes across 21 ARs [73]. Recently, the AR reconnaissance in the northeast Pacific [74] aimed to improve the science and forecasts of landfalling ARs to help better inform decision-makers on water management and flooding hazard in the Western United States. To achieve this goal, NWP models will be combined with observations collected within ARs' core from drop-sondes (deployed by aircraft), and also on the ocean surface (drifting ocean buoys), and airborne radio occultation [75, 76].

In recent years, many studies have highlighted the significance of ARs in shaping the global water vapor distribution, water and energy budgets, and hydrology extremes. Since ARs can affect the Earth's climate through their effect on the poleward transport of water vapor, it is crucial to accurately represent their associated dynamics, thermodynamics, and hydrodynamics in climate models [31, 77]. Climate models to date represent the observed statistics of ARs relatively well, while significant regional biases still exist [77]. An accurate simulation of ARs is crucial to pursue robust projections of AR changes under a warming climate. However, climate models are complex and often bear some degree of uncertainty, and incorporating the fundamental dynamical processes of ARs in a climate model is still under development, and this hindered our understanding of the AR response to a warming climate [31]. Zhang et al. [31] elaborated an idealized atmospheric general circulation model in which an Earth-like global circulation was combined with a hydrological cycle model. The model used passive tracers, simplified cloud microphysics, and precipitation to model water vapor and clouds. They found a good representation of observed dynamical structures for individual ARs, statistical characteristics of ARs, and spatial distributions of AR climatology.

For an AR forecast to be of interest to the following forecast users' properties at seasonal time scales, late medium range and early extended range and the short and early medium range (1–7 days) [76] should be considered depending on the region of interest.

- Good knowledge of the ARs frequency and the amount of total precipitation induced by ARs within a season across a given region is valuable information that can help decision-makers to manage the consequences of ARs. In the polar region, we know that ARs effects are likely critical during the summer melting season.

- In the late medium range and early extended-range forecast (2–3 weeks), the approximate time when an area may experience extreme precipitation and flooding is desired.

- From a forecast user's perspective, predicting AR in the short and early medium range (1–7 days) is highly important and challenging from a modeling perspective. It is essential to accurately predict the intensity, location, and exact time an AR will make a landfall.

Two newly available AR forecast products are the Extreme Forecast Index (EFI) for water vapor flux [78–81] and the AR Scale [70]. The EFI compares the probability distribution of the European Centre for Medium-Range Weather Forecasts (ECMWF). The climatology of ARs in the polar regions and mountain glaciers is still being studied. There is still a long way to go before developing AR forecasting tools for high-latitude regions.

5. Effect of ARs on sea ice, glaciers, and ice shelves

The importance of ARs has been demonstrated over the Antarctic ice sheet [7, 14, 15, 18, 19, 21, 22, 82–84], the Antarctic sea ice [12], and the Arctic continent [8, 10, 36, 37, 41]. Extreme warmer events usually accompany ARs traveling to Antarctica or Greenland from the midlatitudes. Most of these weather events are formed at lower latitudes, over the Southern Ocean and Atlantic Ocean. They transport heat and moisture, propagate eastward and poleward, dissipate, and lose their baroclinic energy near the continent [36, 85, 86]. ARs generate positive anomalies in temperature, moisture, and winds in the coastal regions bordering the Antarctic and Greenland. Several studies have shown ARs impact both the Antarctic ice sheet and the Greenland Ice Sheet through high snowfall events, surface melt events, and sea ice decays [7, 8, 19, 37, 41, 82–84].

ARs occur in Antarctica on low frequency (around 12 events per year in West Antarctica). They account for the most significant percentage of precipitation observed over Antarctica and have important consequences for ice shelf stability [17, 18, 21]. For example, Wille et al. [82] showed that the exceptional heat wave with widespread 30–40°C temperature anomalies across the ice sheet that occurred over East Antarctica between 15 and 19 March 2022 was triggered by an intense atmospheric river advecting subtropical/midlatitude heat and moisture deep into the Antarctic interior. The March 2022 East Antarctica "Heat" wave caused extensive precipitation and surface melt along coastal areas. They also recorded extensive high snowfall accumulations within the interior of the East Antarctic region that resulted in a primarily positive surface mass balance [83]. Extremes warm events can trigger instantaneous knockout effects on the cryosphere because of positive feedbacks (ice/snow albedo) susceptible to enhance and escalate the warming trend [84]. ARs and the associated warm air intrusion events cause about 3 days of surface melting over the Pine Island Glacier in February and March 2013 [14]. Liang et al. [12] suggested that low-frequency ARs can lead to prominent sea ice reduction over marginal ice zone primarily through thermodynamic processes over the Antarctic sea ice. They also argued that ARs could amplify sea ice melting during cold weather periods in contrast to narrow effects during summertime.

In the Arctic, Li et al. [10] found that individual AR events associated with large cyclones initiate a rapid sea ice decrease through turbulent heat fluxes and winds.

6. AR under future warming

The characteristic of ARs, their number, frequency, and intensity are likely to change in future climates, and it is essential to understand and predict how ARs will change under a warming climate. An increase in atmospheric moisture in a warmer climate will lead to an increase in the magnitude of ARs, with significant consequences for flooding, ice sheet stability, and sea ice cover. There is now a growing number of studies highlighting the importance of investigating the contribution of a globally warming atmosphere and changes in the inflow of relatively warm water when examining future changes in the polar region, especially the surface mass balance over glaciated areas and air-sea interactions over the ice cover region.

AR occurrences for different regions around Antarctica and Greenland have been associated with various teleconnections and modes of natural variability, such as the Southern Annular Mode (SAM, [17]), the Pacific South American Mode 2, the Pacific Decadal Oscillation [87] Arctic Oscillation, North Atlantic Oscillation, and Pacific/North America [88]. We hypothesize that the intensity and frequency of AR events may vary with the SAM and the ENSO. It has been suggested that ENSO events can lead to more blocking in the Amundsen Sea and more robust westerly flow on the continental shelf, while a positive SAM is linked to weaker easterly flow north of the ASE [89]. Recently, [60] reported that the Amundsen Sea blocking activity and a negative SAM correlate with ENSO conditions in the tropical Pacific Ocean during the peak summer warming (December–January) in West Antarctica. Moreover, Shields et al. [25] argued that the Indian Ocean Dipole teleconnections in phase with ENSO produce a stronger AR precipitation response in Antarctica compared to other modes of natural variability that affect the Antarctic continent.

The AR Recon program focus on the North Atlantic and represents a recent initiative to improve our understanding to achieve better forecasts of ARs and their impacts [76]. The program targets to develop advanced numerical methods and observation campaigns, assimilate the observed data, and theoretical physical studies to investigate ARs responses to climate change in Europe and the United States of America. A similar program will be needed for an in-depth study of polar ARs. However, the polar regions are characterized by a range of extreme and continuously harsh environmental conditions, and observation campaigns within polar ARs are challenging.

7. Summary points and future issues

The critical role of ARs in various extreme precipitation and flooding events in Western North America, the Western Pacific, Europe, New Zealand, and South America has been demonstrated in multiple studies over the last two decades. In the polar regions, the effects of ARs have been uncovered in the last decade. ARs are less frequent in the polar regions than in the midlatitudes. However, ARs are shown to have a high impact on the Antarctic Ice Sheet mass balance, as they have been linked to surface melting on the West Antarctic Ice Sheet, record high temperatures, extreme snowfall events on the Antarctic Peninsular and the East Antarctic ice sheet, and sea ice decline in marginal ice zones. In the Arctic, ARs have been shown to affect warming in the troposphere during summertime, and the surface radiative balance with potential consequences for Arctic sea ice [9, 10]. A good understanding of the dynamics that drive ARs is vital for predicting how weather extremes in the Arctic and Antarctic and their impacts will change in response to climate change in those regions.

In this work, we have reviewed the impact of AR on sea ice, mountain glaciers, and glaciers in the polar regions. We have also reviewed recent advances in AR, including the identification methods, their main climatological characteristics, their impacts on glaciers, their relationship with large-scale ocean-atmosphere dynamics, and some variabilities under future climate. Significant progress has been made toward a better understanding of ARs, including addressing uncertainty in the tracking and identification algorithm through the ARTMIP project, a quantitative definition of ARs, observations campaign (the AR Recon program), and modeling efforts for better forecasts of ARs (the EFI for water vapor flux and the AR Scale). However, several questions remain about the mechanisms driving atmospheric rivers and their life cycles (genesis, development, and dissipation), observations of their development, their interaction with large-scale dynamics, their role in ephemeral, extent melt events over the Arctic and Antarctica, forecast ARs accurately using NWP systems to provide warnings and awareness, and a better understanding of how ARs interact with large-scale circulation in a warming climate.

As high latitude is sensitive to warmer temperatures, it is imperative to understand how extreme weather or climate conditions will influence moisture intrusions on the Antarctic continent and Arctic and mountain glaciers. Improving our understanding of the exact mechanisms (thermodynamic or dynamical changes) that underlay the long-term changes in moisture transport associated with ARs remains uncertain in the pair region. Moreover, fields campaign to collect observations within ARs' core in the polar region will be necessary. A better prediction of the influence of anthropogenic changes in the climate of Antarctica and Greenland will require an accurate understanding of the processes and impacts of extreme temperature events associated with atmospheric rivers in these regions.

Conflict of interest

The authors declare no competing interests.

Author details

Georges Djoumna[1*] and Sebastian H. Mernild[2]

1 École Secondaire David-Saint-Jacques, Viamonde School Board, Kitchener, ON, Canada

2 SDU Climate Cluster, University of Southern Denmark, Odense, Denmark

*Address all correspondence to: gdjoumna@gmail.com

References

[1] Ralph MF, Dettinger MD, Cairns MM, Galarneau TJ, Eylander J. Defining atmospheric river: How the glossary of meteorology helped resolve a debate. Bulletin of the American Meteorological Society. 2018;**99**:837-839. DOI: 10.1175/BAMS-D-17-0157.1

[2] Ralph FM, Dettinger M, Lavers D, Gorodetskaya IV, Martin A, Viale M, et al. Atmospheric rivers emerge as a global science and applications focus. Bulletin of the American Meteorological Society. 2017;**98**:1969-1973. DOI: 10.1175/BAMS-D-16-0262.1

[3] Guan B, Waliser DE, Ralph FM. Global application of the atmospheric river scale. Journal of Geophysical Research: Atmospheres. 2023;**128**: e2022JD037180. DOI: 10.1029/2022JD037180

[4] Payne AE, Demory M-E, Leung LR, Ramos AM, Shields CA, Rutz JJ, et al. Responses and impacts of atmospheric rivers to climate change. Nature Reviews Earth & Environment. 2020;**1**:143-157. DOI: 10.1038/s43017-020-0030-5

[5] O'Brien TA, Ashley E, Payne C, Shields A, Rutz J, Brands S, et al. Detection uncertainty matters for understanding atmospheric rivers. Bulletin of the American Meteorological Society. 2020;**101**(6):E790-E796. DOI: 10.1175/BAMS-D-19-0348.1

[6] Zhang Z, Ralph FM, Zheng M. The relationship between extratropical cyclone strength and atmospheric river intensity and position. Geophysical Research Letters. 2019;**46**:1814-1823. DOI: 10.1029/2018GL079071

[7] Gorodetskaya IV, Tsukernik M, Claes K, Ralph MF, Neff WD, Van Lipzig NP. The role of atmospheric rivers in anomalous snow accumulation in East Antarctica. Geophysical Research Letters. 2014;**44**:6199-6206. DOI: 10.1002/2014GL060881

[8] Mattingly KS, Mote TL, Fettweis X. Atmospheric river impacts on Greenland ice sheet surface mass balance. Journal of Geophysical Research: Atmospheres. 2018;**123**(16):85388560. DOI: 10.1029/2018JD028714

[9] Fritts R. Rivers in the sky are hindering winter Arctic Sea ice recovery. Eos. 2023;**104**. DOI: 10.1029/2023EO230098. Published on 13 March 2023

[10] Li L, Cannon F, Mazloff MR, Aneesh C, Subramanian S, Wilson AM, et al. Impact of atmospheric rivers on Arctic Sea ice variations. The Cryosphere. 2024;**18**:121-137. DOI: 10.5194/tc-18-121-2024

[11] Maclennan ML, Lenaerts JTM, Shields C, Wille JD. Contribution of atmospheric rivers to Antarctic precipitation. Geophysical Research Letters. 2022;**49**:e2022GL100585. DOI: 10.1029/2022GL100585

[12] Liang K, Wang J, Luo H, Yang Q. The role of atmospheric rivers in Antarctic sea ice variations. Geophysical Research Letters. 2023;**50**: e2022GL102588, 10.1029/2022GL102588

[13] Zhang P, Chen G, Ting M, et al. More frequent atmospheric rivers slow the seasonal recovery of Arctic sea ice. Nature Climate Change. 2023;**13**:266273. DOI: 10.1038/s41558-023-01599-3

[14] Djoumna G, Holland DM. Atmospheric rivers, warm air intrusions, and surface radiation balance in the

Amundsen Sea Embayment. Journal of Geophysical Research: Atmospheres. 2021;**126**(13):e2020JD034119

[15] Adusumilli S, Fish M, Fricker HA, Medley B. Atmospheric river precipitation contributed to rapid increases in surface height of the West Antarctic Ice Sheet in 2019. Geophysical Research Letters. 2021;**48**: e2020GL091076. DOI: 10.1029/2020GL091076

[16] Terpstra A, Gorodetskaya IV, Sodemann H. Linking sub-tropical evaporation and extreme precipitation over East Antarctica: An atmospheric river case study. Journal of Geophysical Research: Atmospheres. 2021;**126**: e2020JD033617. DOI: 10.1029/2020JD033617

[17] Wille JD, Favier V, Gorodetskaya IV, Agosta C, Kittel C, Beeman JC, et al. Antarctic atmospheric river climatology and precipitation impacts. Journal of Geophysical Research: Atmospheres. 2021;**126**:e2020JD033788. DOI: 10.1029/2020JD033788

[18] Turner J, Lu H, King JC, Carpentier S, Lazzara M, Phillips T, et al. An extreme high temperature event in coastal East Antarctica associated with an atmospheric river and record summer downslope winds. Geophysical Research Letters. 2022;**49**:e2021GL097108. DOI: 10.1029/2021GL097108

[19] Wille JD, Favier V, Jourdain NC, Kittel C, Turton JV, Agosta C, et al. Intense atmospheric rivers can weaken ice shelf stability at the Antarctic Peninsula. Communications Earth & Environment. 2022;**3**:90. DOI: 10.1038/s43247-022-00422-9

[20] Clem KR, Bozkurt D, Kennett D, King JC, Turner J. Central tropical Pacific convection drives extreme high temperatures and surface melt on the Larsen C Ice Shelf, Antarctic Peninsula. Nature Communications. 2022;**13**:3906. DOI: 10.1038/s41467-022-31119-4

[21] Wille JD, Favier V, Dufour A, Gorodetskaya IV, Turner J, Agosta C, et al. West Antarctic surface melt triggered by atmospheric rivers. Nature Geoscience. 2019;**12**:111. DOI: 10.1038/s41561-019-0460-1

[22] Francis D, Mattingly KS, Lhermitte S, Temimi M, Heil P. Atmospheric extremes caused high oceanward sea surface slope triggering the biggest calving event in more than 50 years at the Amery Ice Shelf. The Cryosphere. 2021;**15**:21472165, 2021. DOI: 10.5194/tc-15-2147-2021

[23] Ralph FM, Neiman PJ, Wick GA, Gutman SI, Dettinger MD, Cayan DR, et al. Flooding on Californias Russian River: Role of atmospheric rivers. Geophysical Research Letters. 2006;**33**: L13801. DOI: 10.1029/2006GL026689

[24] Gimeni L, Nieto R, Vázquez M, Lavers DA. Atmospheric rivers: A mini-review. Atmospheric Science. Vol. 2. No. 2. Frontiers in Earth Science; 2014. DOI: 10.3389/feart.2014.00002

[25] Shields CA, Wille JD, Marquardt Collow AB, Maclennan M, Gorodetskaya IV. Evaluating uncertainty and modes of variability for Antarctic atmospheric rivers. Geophysical Research Letters. 2022;**49**: e2022GL099577. DOI: 10.1029/2022GL099577

[26] Dare HF, Martinez-Alvarodo, Mbengue CO. Linking atmospheric Rivers and warm Conveyor Belt airflows. Journal of Hydrometeorology. 2019;**20**: 1183-1196. DOI: 110.1175/JHM-D-18-0175.1

[27] Corringham TW, McCarthy J, Shulgina T, Gershunov A, Cayan DR, Ralph FM. Climate change contributions to future atmospheric river flood damages in the western United States. Nature Scientific Reports. 2022;**12**:13747. DOI: 10.1038/s41598-022-15474-2

[28] Wang S, Ma X, Zhou S, Wu L, Wang H, Tang Z, et al. Extreme atmospheric rivers in a warming climate. Nature Communications. 2023;**14**:3219. DOI: 10.1038/s41467-023-38980-x

[29] Shields CA, Kiehl JT. Atmospheric river landfall-latitude changes in future climate simulations. Geophysical Research Letters. 2016;**43**:8775-8782

[30] Dettinger MD, Ralph FM, Das T, Neiman PJ, Cayan DR. Atmospheric rivers, floods and the water resources of California. Water. 2011;**3**:445-478

[31] Zhang P, Chen G, Ma W, Ming Y, Wu Z. Robust atmospheric river response to global warming in idealized and comprehensive climate models. Journal of Climate. 2021;**34**:77177734. DOI: 10.1175/JCLI-D-20-1005.1

[32] Espinoza V, Waliser DE, Guan B, Lavers DA, Ralph FM. Global analysis of climate change projection effects on atmospheric rivers. Geophysical Research Letters. 2018;**45**:4299-4308. DOI: 10.1029/2017GL076968

[33] Michaelis AC, Gershunov A, Weyant A, Fish MA, Shulgina T, Ralph FM. Atmospheric river precipitation enhanced by climate change: A case study of the storm that contributed to California's Oroville Dam crisis. Earth's Future. 2022;**10**: e2021EF002537. DOI: 10.1029/ 2021EF002537

[34] Sodemann H, Stohl A. Moisture origin and meridional transport in atmospheric rivers and their association with multiple cyclones. Monthly Weather Review. 2013;**141**(8): 2850-2868. DOI: 10.1175/mwr-d-12-00256.1

[35] Bao J-W, Michelson SA, Neiman P, Ralph FM, Wilczak JM. Interpretation of enhanced integrated water vapor bands associated with extratropical cyclones: Their formation and connection to tropical moisture. Monthly Weather Review. 2006;**134**:4. DOI: 10.1175/ mwr3123.1

[36] Nash D, Waliser D, Guan B, Ye H, Ralph FM. The role of atmospheric rivers in extratropical and polar hydroclimate. Journal of Geophysical Research: Atmospheres. 2018;**123**:6804-6821. DOI: 10.1029/2017jd028130

[37] Mattingly KS, Mote TL, Fettweis X, van As D, Van Tricht K, Stef Lhermitte S, et al. Strong summer atmospheric rivers trigger Greenland ice sheet melt through spatially varying surface energy balance and cloud regimes. Journal of Climate. 2020; **33**(16):6809-6832. DOI: 10.1175/JCLI-D-19-0835.1

[38] Simpkins G. Atmospheric river impacts on sea ice. Nature Reviews Earth & Environment. 2023;**4**:295. DOI: 10.1038/s43017-023-00434-9

[39] Bozkurt D, Rondanelli R, Marin JC, Garreaud R. Foehn event triggered by an atmospheric river underlies record setting temperature along continental Antarctica. Journal of Geophysical Research: Atmospheres. 2018;**123**: 3871-3892. DOI: 10.1002/2017JD027796

[40] Nicolas JP, Vogelmann AM, Scott RC, Wilson AB, Cadeddu MP, Bromwich DH, et al. January 2016 extensive summer melt in West

Antarctica favoured by strong El Nino. Nature Communication. 2016;**8**(15799): 1-10. DOI: 10.1038/ncomms15799

[41] Mattingly KS, Turton JV, Wille JD, et al. Author correction: Increasing extreme melt in Northeast Greenland linked to foehn winds and atmospheric rivers. Nature Communications. 2024;**15**: 235. DOI: 10.1038/s41467-023-44592-2

[42] Hansen CS, Underwood SJ. Synoptic-scale weather patterns and large slab avalanches at Mt. Shasta, California. Northwest Science. 2012;**86**: 329-341. DOI: 10.3955/046.086.0408

[43] Hatchett BJ, Burak S, Rutz JJ, Oakley NS, Bair EH, Kaplan ML. Avalanche fatalities during Atmospheric River events in the Western United States. Journal of Hydrometeorology. 2018;**18**:1359-1374. DOI: 10.1175/JHM-D-16-0219.1

[44] Nyland D, Nodelman JR. British Columbia. In: Palko K, Lemmen DS, editors. Climate Risks and Adaptation Practices for the Canadian Transportation Sector 2016. Ottawa, ON: Government of Canada; 2017. pp. 66-103

[45] Rutllanta JA, Matusb F, Rudloffa V, Rondanellib R. The role of atmospheric rivers in rainfall-induced landslides: A study from the Elqui valley. Journal of Arid Environments. 2023;**216**:1-5. DOI: 10.1016/j.jaridenv.2023.105016

[46] Thapa K, Endreny TA, Ferguson CR. Atmospheric rivers carry non- monsoon extreme precipitation into Nepal. Journal of Geophysical Research: Atmospheres. 2018;**123**:59015912. DOI: 10.1029/2017JD027626

[47] Nash D, Carvalho LMV, Rutz JJ, Jones C. Influence of the freezing level on atmospheric rivers in High Mountain Asia: WRF case studies of orographic precipitation extremes. Climate Dynamics. 2024;**62**:589607. DOI: 10.1007/s00382-023-06929-x

[48] Nash D, Carvalho LMV, Rutz JJ, Ding Q. Winter and spring atmospheric rivers in High Mountain Asia: Climatology, dynamics, and variability. Climate Dynamics. 2021;**62**:589607. DOI: 10.1007/s00382-021-06008-z

[49] Sepúlveda SA, Ward BC, Cosman SB, Jacobs R. Preliminary investigations of ground failures triggered during the mid-November 2021 atmospheric river event along the southwestern British Columbia highway corridors. Canadian Geotechnical Journal. 2022;**60**:4. DOI: 10.1139/cgj-2022-009

[50] Mahto SS, Nayak MA, Letternmaier DP, Mishra V. Atmospheric rivers that make landfall in India are associated with flooding. Communication Earth & Environment. 2023;**4**:120. DOI: 10.1038/s43247-023-00775-9

[51] Welker C, Martius O, Froidevaux P, Reijmer CH, Fischer H. A climatological analysis of high-precipitation events in Dronning Maud Land, Antarctica, and associated large-scale atmospheric conditions. Journal of Geophysical Research – Atmospheres. 2014;**119**: 11932-11954. DOI: 10.1002/2014JD022259

[52] Kurita N, Hirasawa N, Koga S, Matsushita J, Steen-Larsen HC, Masson-Delmotte V, et al. Influence of large-scale atmospheric circulation on marine air intrusion toward the East Antarctic coast. Geophysical Research Letters. 2016;**43**:9298-9305 https://doi/10.1002/2016GL070246

[53] Connolley W, M. & King J. C. Atmospheric water vapour transport to

Antarctica inferred from radiosonde data. Quarterly Journal of the Royal Meteorological Society. 1993;**1993**(119): 325342. DOI: 10.1002/qj.49711951006

[54] Naithani J, Galle H, Schayes G. Marine air intrusion into the Adelie Land sector of East Antarctica: A study using the regional climate model (MAR). Journal of Geophysical Research. 2002; **107**(D11):ACL6-1-ACL6-16. DOI: 10.1029/2000JD000274

[55] Nicolas JP, Bromwich D. Climate of West Antarctica and influence of marine air intrusions. Journal of Climate. 2011; **24**(1):49-67

[56] Neff W, Compo GP, Ralph RM, Shupe MD. Continental heat anomalies and the extreme melting of the Greenland ice surface in 2012 and 1889. Journal of Geophysical Research – Atmospheres. 2014;**119**(11):6520-6536. DOI: 10.1002/2014JD021470

[57] Neff W. Atmospheric rivers melt Greenland. Nature Climate Change. 2018;**8**. DOI: 10.1038/s41558-018-0285-8

[58] Bell R, Chu W, Kingslake J, Das I, Tedesco M, Tinto KJ, et al. Antarctic ice shelf potentially stabilized by export of meltwater in surface river. Nature. 2017;**544**:344-348. DOI: 10.1038/nature22048

[59] Zwally HJ, Abdalati W, Herring T, Larson K, Saba J, Steffn K. Surface melt-induced acceleration of Greenland ice sheet ow. Science. 2002;**297**:218-222. DOI: 10.1126/science.1072708

[60] Scott RC, Nicolas JP, Bromwich DH, Norris JR, Lubin D. Meteorological drivers and large-scale climate forcing of West Antarctic surface melt. Journal of Climate. 2019;**32**(2):665-684. DOI: 10.1175/JCLI-D-18-0233.1

[61] Wise MG, Dowdeswell JA, Jakobsson M, Larter RD. Evidence of marine ice-clif instability in Pine island Bay from iceberg-keel plough marks. Nature. 2017;**550**:506-510. DOI: 10.1038/nature24458

[62] Deb P, Orr A, Bromwich DH, Nicolas JP, Turner J, Hosking JS. Summer drivers of atmospheric variability affecting ice shelf thinning in the Amundsen Sea Embayment, West Antarctica. Geophysical Research Letters. 2018;**45**:4124-4133. DOI: 10.1029/2018GL077092

[63] Clem KR, Orr A, Pope JO. The Springtime Influence of Natural Tropical Pacifc Variability on the Surface Climate of the Ross Ice. (Shelf, West Antarctica: Implications for Ice Shelf Thinning) Scientific Reports. 2018;**8**:11983. DOI: 10.1038/s41598-018-30496-5

[64] Shepherd TG. Atmospheric circulation as a source of uncertainty in climate change projections. Nature Geoscience. 2014;**7**:703-708

[65] Shields CA, Rutz JJ, Leung L-Y, Ralph FM, Wehner M, Kawzenuk B, et al. Atmospheric river tracking method intercomparison project (artmip): Project goals and experimental design. Geoscientic Model Development. 2018; **11**(6):24552474

[66] Rutz JJ, Shields CA, Lora JM, Payne AE, Guan B, Ullrich P, et al. The atmospheric river tracking method intercomparison project (ARTMIP): Quantifying uncertainties in atmospheric river climatology. Journal of Geophysical Research: Atmospheres. 2019;**124**(21):13777-13802. DOI: 10.1029/2019JD030936

[67] Cordeira JM, Ralph FM, Martin A, Gaggini N, Spackman JR, Neiman PJ, et al. Forecasting atmospheric Rivers

during CalWater 2015. Bulletin of the American Meteorological Society. 2017; **98**:449-459. DOI: 10.1175/BAMS-D-15-00245.1

[68] Muszynski G, Kashinath K, Kurlin V, Wehner M. Topological data analysis and machine learning for recognizing atmospheric river patterns in large climate datasets. Geoscientific Model Development. 2019;**12**:613628. DOI: 10.5194/gmd-12-613-2019

[69] Xu G, Wang L, Chang P, Ma X, Wang S. Improving the understanding of atmospheric river water vapor transport using a three-dimensional straightened composite analysis. JGR Atmosphere. 2022;**127**(11). DOI: 10.1029/2021JD036159

[70] Ralph MF, Rutz JJ, Cordeira JM, Dettinger M, Anderson M, Reynolds D, et al. A scale to characterize the strength and impacts of atmospheric rivers. Bulletin of the American Meteorological Society. 2019;**100**(2):269289. DOI: 10.1175/BAMS-D-18-0023.1

[71] Wille JD, Bromwich D, Cassano JJ, Nigro MA, Mateling ME, Lazzara MA. Evaluation of the AMPS boundary layer simulations on the Ross Ice Shelf, Antarctica, with unmanned aircraft observations. Journal of Applied Meteorology and Climatology. 2017; **56**(8):22392258. DOI: 10.1175/JAMC-D-16-0339.1

[72] Gorodetskaya IV, Silva T, Schmithsen H, Hirasawa N. Atmospheric river signatures in radiosonde profiles and reanalyses at the Dronning Maud land coast, East Antarctica. Advances in Atmospheric Sciences. 2020;**37**(5): 455476. DOI: 10.1007/s00376-020-9221-8

[73] Ralph FM, Iacobellis S, Neiman P, Cordeira J, Spackman J, Waliser D, et al. Dropsonde observations of total integrated water vapor transport within North Pacific atmospheric Rivers. Journal of Hydrometeorology. 2017;**18**: 2577-2596. DOI: 10.1175/JHM-D-17-0036.1

[74] Ralph FM et al. West coast forecast challenges and development of atmospheric river reconnaissance. Bulletin of the American Meteorological Society. 2020;**101**:E1357-E1377. DOI: 10.1175/BAMS-D-19-0183.1

[75] Lavers DA, Ralph FM, Richardson DS, Pappenberger F. Improved forecasts of atmospheric rivers through systematic reconnaissance, better modelling, and insights on conversion of rain to flooding. Communication Earth and Environment. 2020. DOI: 10.1038/s43247-020-00042-1

[76] Lavers DA, Wilson AM, Ralph FM, Tallapragada V, Pappenberger F, Reynolds C, et al. Advancing Atmospheric River science and inspiring future development of the Atmospheric River reconnaissance program. Bulletin of the American Meteorological Society. 2024;**105**:E75-E83. DOI: 10.1175/BAMS-D-23-0278.1

[77] Waliser DE, Cordeira JM. Atmospheric river modeling: Forecasts, climate simulations, and climate projections. In: Ralph FM et al., editors. Atmospheric Rivers. Switzerland AG: Springer; 2020. pp. 179-199

[78] Lavers DA, Pappenberger F, Richardson DS, Zsoter E. ECMWF extreme forecast index for water vapor transport: A forecast tool for atmospheric rivers and extreme precipitation. Geophysical Research Letters. 2016;**43**:11852-11858

[79] Lavers DA, Zsoter E, Richardson DS, Pappenberger F. An assessment of the ECMWF extreme forecast index for

water vapor transport during boreal winter. Weather and Forecasting. 2017; **32**:1667-1674

[80] Lavers DA et al. Earlier awareness of extreme winter precipitation across the western Iberian Peninsula. Meteorological Applications. 2018;**25**: 622-628

[81] Lavers D, Tsonevsky I, Richardson D, Pappenberger F. The extreme forecast index for water vapour flux. ECMWF Newsletter. 2019;**160**:4-5

[82] Wille JD et al. The extraordinary March 2022 East Antarctica heat wave. Part 1: Observations and meteorological drivers. Journal of Climate. 2024;**37**: 757-778

[83] Wille JD et al. The extraordinary March 2022 East Antarctica heat wave. Part II: Impacts on the Antarctic Ice Sheet. Journal of Climate. 2024;**37**(3): 779-799. DOI: 10.1175/JCLI-D-23-0176.1

[84] Gorodetskaya IV et al. Record-high Antarctic Peninsula temperatures and surface melt in February 2022: A compound event with an intense atmospheric river. npj Climate and Atmospheric Science. 2023;**6**:202. DOI: 10.1038/s41612-023-00529-6

[85] Uotila P, Vihma T, Tsukernik M. Close interactions between the Antarctic cyclone budget and large-scale atmospheric circulation. Geophysical Research Letters. 2013;**40**:3237-3241

[86] Papritz L et al. The role of extratropical cyclones and fronts for Southern Ocean freshwater fluxes. Journal of Climate. 2014;**27**:6205-6224

[87] Turner J, Marshall GJ, Clem K, Colwell S, Phillips T, Lu H. Antarctic temperature variability and change from station data. International Journal of Climatology. 2019;**40**:2986-3007. DOI: 10.1002/joc.6378

[88] Zhang C et al. Climatology and decadal changes of Arctic atmospheric rivers based on ERA5 and MERRA-2. Environmental Research: Climate. 2023; **2**:035005

[89] Turner J, Orr A, Gudmundsson GH, Jenkins A, Bingham RG, Hillenbrand C-D, et al. Atmosphere- Ocean-ice interactions in the Amundsen Sea Embayment, West Antarctica. Reviews of Geophysics. 2017; **55**:235-276. DOI: 10.1002/2016RG000532

Section 2

Methodology and Interpretation of Data

Chapter 2

Drilling Ice and Subglacial Rock Cores for Scientific Discovery in a Changing Climate

Mary R. Albert, Kristina Slawny, Jay Johnson, Elliot Moravec and Tanner Kuhl

Abstract

Ice cores drilled from glaciers and ice sheets provide a critical natural archive of current and past evidence of climate and environmental change, and subglacial rock holds evidence of past glacial extent. Current climate change is causing the demise of glaciers around the world; the scientific need to recover ice cores from mid-latitude glaciers is urgent before ice core records are lost to melt. Logistical access to uncertain ice sheet conditions is challenging. Retrieval of subglacial rock cores is needed for cosmogenic dating evidence of past sea level. This paper describes recent engineering advances in scientific drilling of ice and subglacial rock cores under conditions of current climate change. The successful efforts of the U.S. Ice Drilling Program to retrieve a surface-to-bedrock ice core from Quelccaya Ice Cap, Peru is described, along with the successful subglacial rock coring that retrieved the first meter-length bedrock cores underlying 509 meters of the Greenland Ice Sheet.

Keywords: ice core, bedrock core, subglacial, scientific drilling, glaciers, ice sheets

1. Introduction

The world's glaciers and ice sheets are huge ice masses that have important effects on sea level, water supply, weather and climate; in addition, they serve as natural archives of evidence of past environment and climate due to the depositional nature of their formation. In high altitude and high latitude regions of the Earth, existing glaciers and ice sheets began their formation many thousands to several millions of years ago as snow in places so cold that snow melt rarely occurred. Over many years of snow accumulation, the weight of additional snow and metamorphic processes compacted the underlying snow into firn (old snow). Over time the firn further compacted and transformed into solid glacial ice. Due to glacial flow, the basal area of the glacier may have entrained sand or till. The bed of the glacier may be sediment or till that rests on bedrock of the landform. A schematic of a cross-section of a glacier is shown in **Figure 1**.

In addition to many individual alpine glaciers that exist in high-altitude sites worldwide, Earth has two enormous polar ice sheets that each cover millions of

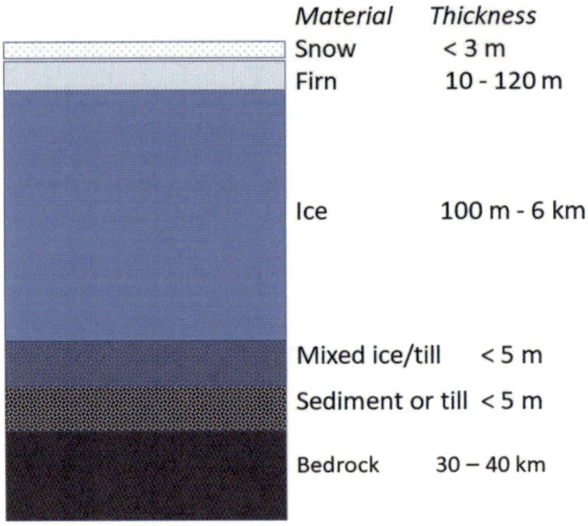

Figure 1.
Schematic illustration of a glacier or ice sheet cross section. Recent snow on the surface accumulates on older snow (firn) that becomes compressed into glacial ice. Basal ice may be mixed with sediment or till, and there may be wet or dry sediment, till, or rocks on the bedrock. Subglacial lakes are not depicted here, and the schematic is not to scale. (credit: IDP).

square kilometers of land: the Greenland Ice Sheet, which is now more than 3 km thick in the center, and the Antarctic Ice Sheet, that is as much as 6.4 km thick in some places. Both ice sheets have many named glaciers that slowly drain ice toward the coast. Scientific discoveries based on evidence of past climate that is embedded in glaciers and ice sheets have revolutionized the field of climate science and enabled understanding of current climate conditions in the context of recent and distant past climate.

Science and engineering are inextricably linked in glacial endeavors. Retrieval of ice cores from glaciers and ice sheets requires highly specialized ice drilling engineering. While many nations have institutions that conduct ice drilling, the U.S. Ice Drilling Program (IDP) is funded by the National Science Foundation to lead integrated planning for ice drilling and coring and to provide the expertise and drilling technology needed to advance ice core and subglacial science. This chapter briefly describes the nature of climate evidence held within and beneath glaciers, scientific drivers of several current societally-relevant projects, and engineering endeavors by IDP that enable discoveries even as the world's glaciers and ice sheets are changing under current climate warming.

2. Glaciers: an archive of evidence about environment and climate

In cold, high-latitude and high-altitude regions where snow does not melt, annual snowfall can accumulate over hundreds of thousands of years on the polar ice sheets. The snow crystals contain water isotopes and chemical impurities that hold clues to atmospheric conditions of their origin. As recent snow accumulates, the underlying snow from years and decades in the past, called firn, becomes compressed and undergoes post-depositional metamorphism over depths that range from 10 to 120 meters. In so doing, the interstitial space between the snow crystals becomes constricted until the remaining air finally becomes trapped in bubbles in solid ice; the air in the bubbles contains direct evidence of atmospheric composition. Over many thousands of years of accumulation, the glacial ice can become multiple kilometers thick. The ice and trapped gases near the bottom hold the oldest records while the snow at the top holds the youngest.

The accumulation process that formed glaciers and ice sheets also archives a record of environmental conditions. For example, stable isotopes of oxygen and hydrogen (oxygen-18 and deuterium) in an ice core were initially deposited as snow, and are indicators of past temperature variations. Moisture in the atmosphere has an isotopic composition related to the vapor source and transport pathways. As winds transport the moisture to an ice sheet, temperature-dependent isotopic fractionation occurs where heavier isotopes preferentially fall out as precipitation. The moisture that is deposited onto the ice sheet as snow eventually becomes archived as glacial ice, and the isotopes provide clues to the transport process as well as the temperature when the snow fell. In addition, chemical composition in snowfall provides a wealth of evidence including changes in atmospheric circulation, biological activity, volcanic eruptions, fire, and air pollution for example; such evidence becomes chronologically archived in the ice. The composition of the air in the bubbles provides direct evidence of past atmospheric composition, for example levels of carbon dioxide, methane, and other greenhouse gases. A description of stable water isotopic fractionation with examples of an ice core climate record from the Vostok ice core in Antarctica are provided in [1] and an overview of common chemical measurements for ice core records is provided in [2].

In consistently very cold locations that have high snow accumulation rates, glacial ice contains high-resolution annual records of past environment and climate that provide a critical past context for comparison with present conditions along with an understanding of Earth systems, including behavior of major atmospheric circulation patterns. For example, records from ice cores drilled at a high-altitude site on Mount Logan, the highest mountain in Canada, show evidence of the regional Hadley and Walker circulations in the Pacific [3], and stable isotopes indicate that shifts between modern and zonal atmospheric flow regimes occurred abruptly within a few years, with some coinciding with the end of the Little Ice Age and the beginning of the European Warm Period [4]. Mount Logan ice core records of soluble sodium, a proxy for the strength of the wintertime Aleutian Low, displays variability that shows strong similarities to tropical proxy records for the El Nino-Southern Oscillation [5].

In locations where snow melt occurs and the meltwater flows through the firn and ice, the chemical and isotopic evidence becomes mixed and blurred, and the natural archive can be destroyed. Scientists have documented surface area decrease in many glaciers, including four glaciers in Peru, Bolivia, equatorial east Africa, equatorial Papua, Indonesia, and the western Tibetan Plateau [6]. In that study the ice core records based on oxygen isotopic ratios (δ18O) show substantial impact of seasonal melt in the firn layers in ice cores from elevations below ~6000 m, however δ18O records recovered from higher altitude sites still contain well-preserved seasonal variations to the surface. Unfortunately, as climate warming continues, higher elevation glaciers will eventually also be degraded.

2.1 Drill before it's gone: ice cores in a warming world

The polar ice sheets and alpine glaciers around the world are experiencing dramatic changes as they melt under current climate warming. According to the IPCC, in the two decades leading up to 2020 Greenland lost approximately 4890 Gt of ice and it is virtually certain that the Greenland Ice Sheet will continue to lose mass over the twenty-first century [7].

Because the high-resolution chemical and isotopic records archived in alpine glaciers provide patterns of regional climate variability over the many centuries to millennia, the demise of glaciers in temperate regions is removing irreplaceable clues to paleo regional climate patterns in the midlatitude regions of the planet. Originally formed during the most recent ice age and receding during the current interglacial cycle, many midlatitude glaciers have now stopped accumulating mass; on those glaciers the snow accumulated during the winter disappears that summer due to melt. Temperatures in the frozen depths of most midlatitude glaciers are gradually warming, with some now approaching the pressure melting point.

Glaciers in the central Andes Mountains of South America archive evidence of mid-latitude climate patterns that have a large impact on weather and human activity in the equatorial region and beyond. The Quelccaya Ice Cap in Peru, the largest remaining tropical ice cap in the western hemisphere, is a remnant of the large ice sheet that covered South America during the Pleistocene era. Quelccaya has had multiple retreats and advances since the Last Glacial Maximum about 20,000 years ago [8]. Near the summit of Quelccaya, ice cores were drilled to bedrock 2003; analysis of a core revealed precipitation patterns, volcanic activity, and climate variability at annual resolution over the past 1800 years along with shifts in the Intertropical Convergence Zone over this region of the Andes [9]. Ice drill engineering enabled a lighter weight drill with low fuel requirements for this remote location [10]. From shallow cores drilled in 2018, it has become evident that warming is causing meltwater percolation through the firn, washing out some climate evidence in the firn and ice near the firn-ice transition; however particle-based trace element records (e.g., Fe, Mg, K) retained well-preserved signals, yielding hope that another deep ice core from Quelccaya may offer additional discoveries [11]. The Quelccaya Ice Cap, which covered an area of approximately 50.2 km2 in 2018, may completely lose its accumulation zone before the end of the twenty-first century or sooner as the ice cap continues on a trajectory to disappearance [12]. Retrieval of a new surface-to-bedrock ice core from the summit region of Quelccaya before the glacier suffers further demise will retrieve glacial ice that can be preserved and used with new high-resolution chemical sampling techniques to enable additional climate discoveries.

Drilling ice cores in warm ice that is ice close to its pressure melting point is challenging. Commonly used cable-suspended electromechanical ice coring drills, which use a rotating cutting head with steel blades that cut an annulus around the core, can become stuck as ice that is close to the pressure melting point around the sonde melts and refreezes. At sites such as Quelccaya, melting an annulus around the core is the preferred coring method. Electrothermal drills are cable-suspended drills that have an electric powered heat ring at the drill head to melt through the ice. Photos of the drill head of the IDP Electrothermal Drill are shown in **Figure 2a** and the drill with an ice core in **Figure 2b**.

Variants of electrothermal drill types have been used by ice coring programs in a number of nations; a summary of the applications can be found in [13]. The current IDP Electrothermal Drill is an upgrade of a thermal drill from the former Polar Ice Coring Office. The drill currently consists of the sonde with heated drill head, which is used in conjunction with the cable, winch and control system of the IDP 4-Inch Drill. An 1800 W (180 V) heat ring is located at the end of the sonde, as well as three spring-loaded blades called core dogs that are recessed in the core barrel wall and

(a)

(b)

Figure 2.
(a) The heat ring serves as the drill head of the IDP Electrothermal drill (photo credit: IDP). (b) an ice core drilled with the Electrothermal drill during the 2023 field work at combatant col, Mt. Waddington, British Columbia, fieldwork (photo credit: Peter Neff).

engaged at core break to hold the core in the barrel until it is unloaded at the surface. The system is used to collect two-meter length, 86 mm diameter, ice cores using power supplied by a 5 kW generator. To remove an ice core from the barrel, the drill is laid horizontally and the core dogs are retracted using three magnets. Disengaging the core dogs allows the core to be pushed out through the heat ring. Two prototype tools have recently been designed by IDP and deployed to assist with difficult drilling conditions. Any debris encountered in the borehole can limit or prevent heat transfer and penetration. A debris vacuum assembly consists of a vacuum head with internal one-way valves attached to a tube containing a spring-loaded piston. A trigger mechanism in the vacuum head releases the piston when the tool is set on the bottom of the borehole, sucking any debris smaller than approximately 3 mm in diameter up into the tube. In conditions where the risk of borehole refreezing is present, an ethanol deployment system consisting of a tube containing a long plastic bag filled with ethanol may be deployed. A sharp arrowhead on the end of a plunger arm is used to puncture the bag and deploy the ethanol at the bottom of the borehole. Successful ice core drilling requires many adaptations that are specific to the nature of the ice. The IDP Electrothermal Drill was used in 2022 by IDP to successfully retrieve an ice core from the surface down to bedrock at 128 m depth at the Quelccaya Ice Cap. **Figure 3** shows the Electrothermal drill with its tent, and **Figure 4** shows drill operations at Quelccaya.

In addition to challenges with the drilling, there are significant logistical constraints on accessing the remote Quelccaya ice cap and retrieving ice cores. Commercial transportation is available to the city of Cusco, Peru. From there, one must contract with local logistics personnel for transport of the field team and equipment to the field site. Car transport moves gear and personnel from Cusco to Phinaya (16,600 ft) at the base of the ice cap. Starting in Phinaya, cargo is loaded onto horses for transport up the ice cap. A basecamp was established at 17,000 ft., at which point porters load and carry cargo to the summit at 18,600 ft., as depicted in **Figure 5**. The slow ascent to the summit of the ice cap requires multiple days of acclimatization along the way to adapt to the high altitude.

Drilling operations at the Quelccaya site must be light and agile by necessity at this remote, high altitude site. The total camp is shown in **Figure 6**. Smaller tents are used for berthing while a larger tent is adapted to house the drill.

Figure 3.
The IDP Electrothermal drill at the drill camp on Quelccaya ice cap (photo credit: Mariusz Potocki).

DOI: http://dx.doi.org/10.5772/intechopen.1004695

(a)

(b)

(c)

Figure 4.
(a) The Electrothermal drill sonde hangs on the drill tower and is lowered into the borehole (photo credit: Mariusz Potocki). (b) IDP engineer Elliot Moravec operates the drill on Quelccaya ice cap (photo credit: Mariusz Potocki). (c) the IDP Electrothermal drill operations are conducted in a Tentipi Safir tent (photo credit: Mariusz Potocki).

Figure 5.
Porters carry drill cargo, ice core boxes and field camp gear through fields of Penitentes ice formations up to Quelccaya ice cap (photo credit: Elliot Moravec).

Figure 6.
Drilling camp on Quelccaya ice cap in 2022 (photo credit: Mariusz Potocki).

Figure 7.
The science team measures and processes an ice core drilled on Quelccaya (photo credit: Mariusz Potocki).

With scant resources at the site, the team measures, weighs and packs the core as it is drilled at the site, in preparation for transporting the core down the glacier. **Figure 7** depicts scientists working with the core at the drill site.

The ingenuity, experience, and hardiness of engineers in the Ice Drilling Program and the Quelccaya field team has achieved the successful retrieval of an irreplaceable ice core from Peru. Chemical clues in the ice cores that may reveal environmental conditions and climate shifts in the midlatitudes pre-dating current anthropogenic climate warming are now stored at laboratories in the U.S., safely archived for science even as the glacier melts away.

3. Paleo sea level: evidence of glacial extent embedded in subglacial rock

Over long periods of time, a balance of water on Earth shifts between oceans and ice sheets on land. In cold glacial periods, ice sheets are huge and sea level is low, while in warm interglacial periods glacial extent is low, and sea level is high. Earth is now in an interglacial period with the Greenland Ice Sheet the only remaining part of the large Laurentide ice sheet that once covered Greenland plus what is now Canada and the northern tier of the U.S. in the previous glacial period. Current anthropogenic climate warming is causing faster melting of glaciers and ice sheets with associated sea level rise. With more than two-thirds of the world's fresh water currently locked up as ice in ice sheets and glaciers, knowledge of past glacial extent can help to understand the future rise in sea level that we are facing.

Rock formations on lands that are now covered by glaciers but that were not covered during the previous interglacial period hold isotopic evidence of local glacial extent in the past, through phenomena associated with their exposure to the sun. Explosions on stars in outer space and on our sun emit cosmic rays, which are high energy particles that flow through space, some of which are constantly hitting the surface of the Earth. Cosmic rays cannot penetrate into a glacier however they can penetrate into the first meter of an exposed rock to create isotopes called cosmogenic nuclides. These radionuclides are produced at a known rate and they undergo radioactive decay at a known rate, so measuring the abundance of the nuclide in a rock sample can indicate how long the rock was exposed to the sun.

3.1 Subglacial geology and past ice extent: recent endeavors in Greenland

The formidable engineering challenges involved in retrieving subglacial bedrock cores have only been overcome very recently. While glaciers are composed of ice for the majority of the hundreds to thousands of meters of their thickness, the basal meters of glaciers are often complex in geometry and composition, sometimes being a mix of glacial ice, sediment, loose rock. There may or may not be cracks and cavities in that mixed region. Below the glacier, the ground may be wet or dry and may have sediment and surface rubble that overlie the bedrock. Retrieving sub-ice bedrock cores is complex and it requires a means of drilling down through the glacier, through mixed ice/rock media near the bottom of the glacier, and then into solid bedrock. A major challenge with drilling through glaciers and into the bedrock is the possibility of hydrofracturing the drilled hole if drilling fluid circulation pressure exceeds the strength of the ice. Drill fluid is circulated during subglacial coring to cool the drill bit and to remove cuttings from the borehole. If hydrofracture occurs, then the drilling fluid drains into the cracks in the ice, drill fluid circulation is lost, and drilling is stopped. Ice is inherently a brittle material, and cracks propagate quickly. Hydrofracture occurs naturally near the surface of ice sheets as surface meltwater streams enter cracks [14]. In a laboratory study, Chen and others [15] found that there is a relationship between the overburden pressure at the depth of fracture and the pressure required to cause fracture, and that small fissures significantly weaken the borehole in phenomena that are temperature dependent. Successful drilling in complex glaciers relies on innovations in engineering.

There have been past efforts by multiple nations to gather subglacial cores through a variety of means; Talalay [16] provides a history of efforts to retrieve samples from the sub-ice environment. When the sub-ice environment is till or soft sediment, ice core drills or hot water drills can be used to create an access hole through the ice, after which metal tubes can be driven into the till or soft sediment to retrieve a core, for example using gravity corers and piston corers. Gravity and piston corers have been successful in till and sediment, however they are unable to retrieve a core from bedrock. On land that is not glaciated, the minerals industry has long drilled rock cores using heavy rotary drilling with a pipe string. Based on that technology, the Rapid Access Ice Drill (RAID) system was designed in order to rapidly drill through thousands of meters on ice sheets [17, 18]. This system will be appropriate for drilling through kilometers of ice as part of a heavy traverse of an ice sheet, however the components are too large to be transported to remote sites by small or midrange aircraft.

Scientific interest in determining past ice extent and sea level often focuses on locations near bedrock outcrops and glacial margins on polar sites where the environment is severe and the logistics are scarce. Sites with limited logistics adds the additional constraint that the drilling system must be agile and portable. Recently there have been engineering breakthroughs in agile ice drilling [19]. Drilling meters of bedrock from beneath many meters of glacial ice using agile, aircraft-deployable drill systems has been pioneered by the U.S. Ice Drilling Program, who developed the ice-enabled Winkie drill [20] and the Agile Sub-Ice Geological Drill [21]. These unique, fully operational drills have been successful in drilling subglacial rock cores from remote sites in Antarctica and also recently in Greenland.

The ice-enabled Winkie drill is an adaptation of the Minex commercial Winkie rock drill, which is a lightweight, compact drill used by the minerals exploration industry to drill small rock cores on unglaciated lands. IDP upgraded various parts of the drill, including replacing the gasoline engine with an electric motor, and also adding ice augering and ice coring capability. The ice-enabled Winkie is capable of drilling through ice, rock, till, and mixed ice/rock systems at locations where there are not cavities or cracks in the ice, nor presence of liquid water. The drill requires pressurized drilling fluid circulation; if there are cracks or cavities in the ice, there would be a loss of fluid in the borehole which would end the drilling. The system can retrieve cores either 33.4 mm or 71.7 mm in diameter and 1.5 m long, and it is able to drill to 120 m depth with the smaller diameter configuration. When there is a significant depth of snow and firn above the glacial ice, a lightweight ice coring drill is used to create a pilot hole through the firn. The system is transportable by helicopter or light aircraft and it can be set up in 3 hours. A schematic of the drill is shown in **Figure 8**. A mixed-media core composed of ice, silt, and rock core that was produced by the ice-enabled Winkie Drill in Ong Valley, Antarctica is depicted in **Figure 9**.

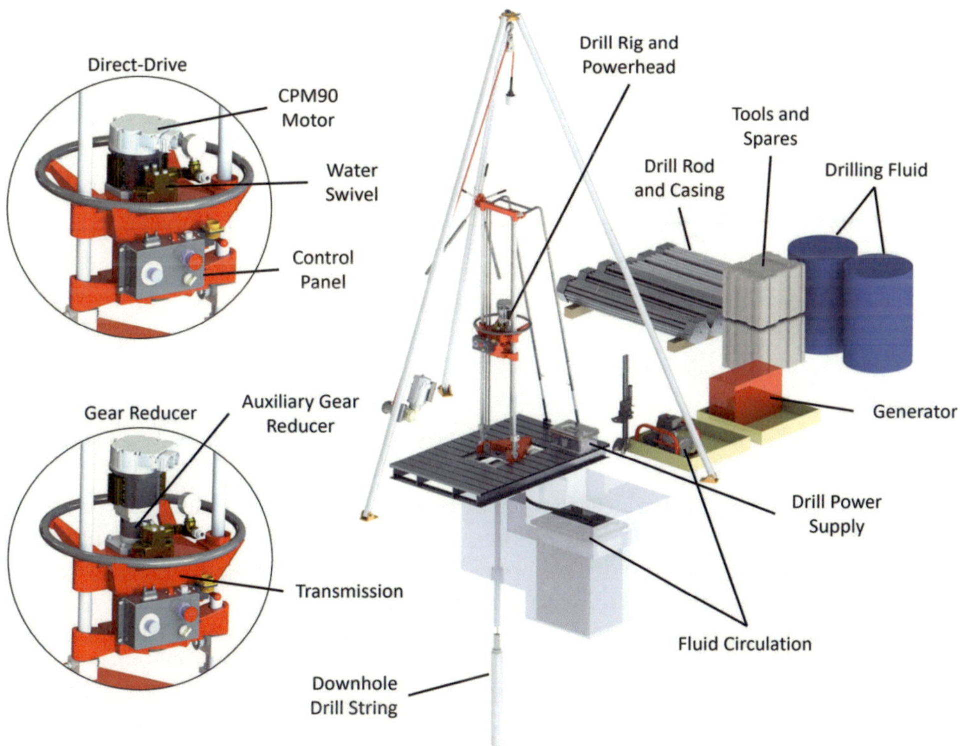

Figure 8.
Schematic of the ice-enabled Winkie drill. (credit: IDP).

Figure 9.
A mixed media core showing ice, silt, and rock that was drilled using the ice-enabled Winkie drill (photo credit: Grant Boeckmann).

Figure 10.
The ice-enabled Winkie drill site at the 2023 GreenDrill project in NW Greenland (photo credit: Elliot Moravec).

The small footprint and relative agility of the ice-enabled Winki drill facilitates its use in drilling through several hundred meters of ice to retrieve meters of subglacial rock core. **Figure 10** shows the drill in operation in northwest Greenland in 2023.

IDP used the ice-enabled Winkie to drill in 2023 in northwest Greenland at a site that is 25 m north of the divide at Prudhoe Dome as part of the GreenDrill project. There the drill performed well through 96 m of ice to retrieve 2 m of frozen subglacial sediment. Twenty days were spent at the Winkie site including setup, takedown, 5.5 days of coring operations, four inactive days due to severe weather, and time spent drilling through the challenging subglacial environment. In total, more than 2500 m of drill rod was tripped in and out of the hole in working to clear and change the bits. A new full face ice bit was used to rapidly penetrate through the glacier ice and access

Figure 11.
Core from the ice-enabled Winkie drilling near Prudhoe dome (photo credit: Jason Briner).

the bed. A new slip foot clamp was designed by IDP and deployed as well as a capstan winch to assist with rod tripping. A new full face drag bit was utilized to quickly penetrate through dirty ice. As the system is operated out in the elements, considerable time was spent digging out the drill site each day and removing snow from the capstan winch, mud pumps and fluid filtration system. Tents to shield components and operations from blowing snow are now being implemented for the next field season. Drilling in severe conditions with cold temperatures and blowing snow, along with scarce logistics requires taking a variety of tools and equipment that may be needed for rapidly changing environmental conditions. **Figure 11** shows a rock core drilled using the ice-enabled Winkie at the site.

While the ice-enabled Winkie drill can be used at shallow sites, drilling through thicker ice with bedrock coring is accomplished using the IDP Agile Sub-Ice Geological Drill (ASIG), which is capable of drilling through 700 m of glacial ice and mixed media to retrieve up to 10 m of bedrock core that is 39 mm in diameter. The IDP ASIG drill is the first proven system in the world with this capability. The development of the ASIG Drill is described in [21]. A schematic of the layout for the ASIG drill is shown in **Figure 12**.

The ASIG Drill is based on a customized Discovery MP1000 Man Portable Diamond Core Drill made by Multi-Power Products Ltd. in Kelowna, BC, Canada. The system uses four Kubota diesel engines, Sandvik WL56 thin-kerf drill rod and Sandvik TK56 wireline coring components. An inflatable packer is used to create a seal between the casing and the ice wall of the borehole, preventing drilling fluid from flowing up into the permeable firn above the packer. An isoparaffinic Naphtha fluid called Isopar K is used as a drilling fluid to transport cuttings to a filtration system on the surface. Polycrystalline diamond compact (PDC) full face bits are used for access drilling. An electronic pressure relief valve (PRV) was implemented by IDP to safeguard against over pressurization of the system and reduce the risk of hydrofracture of the ice. Despite such safeguards, hydrofracture events have occurred

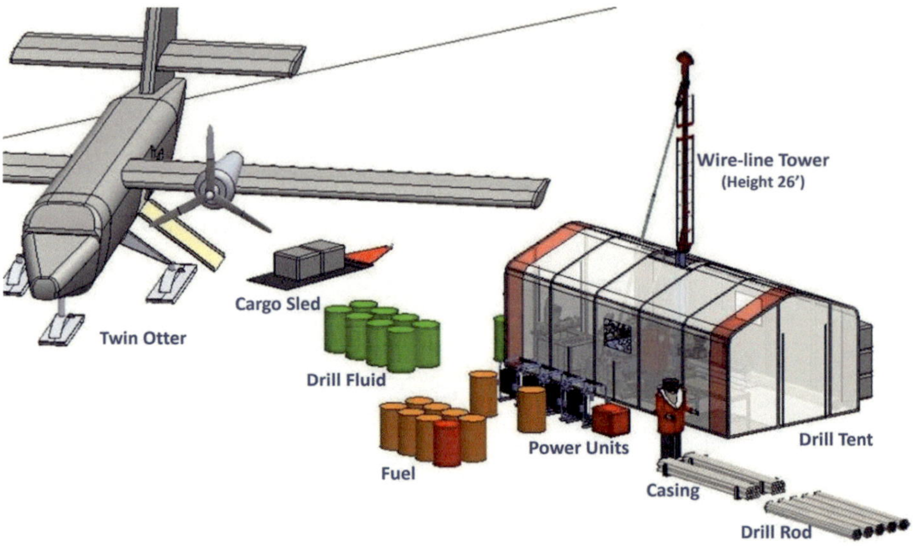

Figure 12.
Layout of the ASIG drill (credit: IDP).

Figure 13.
ASIG camp at Prudhoe dome Greenland in 2023 (photo credit: Tanner Kuhl).

during use of the drill in both Antarctica and Greenland. In Antarctica, the drill was moved and a new hole initiated and completed. In Greenland, IDP engineers were able to remove the casing and packer and deepen the pilot hole several meters using an ice coring drill and then resume drilling.

Both the IDP ASIG and the ice-enabled Winkie drills have been successfully used to retrieve subglacial sediment and rock cores. Briner and others [22] conducted investigations to identify regions of the Greenland Ice Sheet where the ice sheet is less than 700 m thick and the bed of the ice sheet is likely to be frozen with underlying quartz-bearing rock lithologies that are amenable to cosmogenic nuclide analysis. They identified regions in northern, northwestern, and northeastern Greenland that meet the criteria and thus are likely to hold evidence of whether the ice sheet disappeared in past interglacial times 420,000 and 120,000 years ago, along with the rate of melt.

In 2023, IDP used the ASIG Drill in the Prudhoe Dome region of Greenland. Sixty thousand pounds of drilling equipment, drilling fluid, fuel, camping gear and a 10-person team were flown by Basler aircraft and helicopter from Pituffik Space Base in NW Greenland to the field site. Fifty-one days were spent onsite, including 34 days of drilling activity, and 12 days of poor weather that limited drilling, along with several days spent troubleshooting. A hydrofracture occurred just below the packer. The casing and packer were then reset and drilling resumed, resulting in the collection of 3.5 m of subglacial sediment and 4.5 m of gneiss bedrock core from under 509 m of ice. Normal versus reverse fluid circulation reduced borehole pressure at the packer and will likely be used during future ASIG drilling campaigns. The ASIG drill tent is seen at the 2023 Prudhoe Dome drill camp in **Figure 13**.

The tent designed for use with the ASIG drill is relatively lightweight yet provides protection from blowing snow; this facilitates continued drilling during bad weather. The ASIG drill in operation is shown in **Figure 14**. A tray of subglacial rock core retrieved by IDP engineers using the ASIG drill at Prudhoe Dome is shown in **Figure 15**.

Figure 14.
ASIG drill operations in NW Greenland in 2023 (photo credit: Elliot Moravec).

Figure 15.
Subglacial rock core from the ASIG drilling at Prudhoe dome (photo credit: Allie Balter-Kennedy).

4. Conclusions

Glaciers and ice sheets are unique natural archives of current and past evidence of climate and environmental change. Analysis of the evidence in ice core has led to many important discoveries, for example that the climate can change abruptly in less than 10 years [23], and the realization that direct evidence from samples of ancient air in ice cores indicates that levels of carbon dioxide now in our atmosphere are higher than they have been in at least 800,000 years [24, 25]. Retrieval of ice cores from remote locations requires highly specialized engineering, both for development of appropriate drills and for expert use of the drills in a medium as complex as a glacier and the subglacial environment. As current climate change is causing the demise of glaciers around the world and increasing sea level, the need to retrieve scientific evidence from within and beneath glaciers and ice sheets is urgent. The U.S. Ice Drilling Program (IDP) has recently succeeded in ice drilling achievements that are enabling cutting-edge science. Drilling the Quelccaya, Peru ice core from surface to bedrock using the IDP Electrothermal Drill on this remote, high-altitude glacier successfully preserved an ice core that is a natural archive of mid-latitude environment and climate over the past several thousand years, even as the glacier itself is on a trajectory of demise due to melt. In addition, IDP pioneered development of the first proven drilling systems to retrieve rock and sediment cores from beneath hundreds of meters of glacial ice, the Agile Sub-Ice Geological (ASIG) Drill, and the smaller ice-enabled Winkie drill. In 2023, IDP engineers successfully recovered the first meters of rock cores from beneath 509 m of the Greenland Ice Sheet, cores that scientists will analyze to improve understanding of sea level during past interglacial times. The changing climate is creating urgency for new knowledge; engineering efforts of IDP are rising to meet the need, as evidenced by the successful IDP retrieval of the Quelccaya 2023 ice core and the first retrieval of meters of bedrock core from beneath 509 m of the Greenland Ice Sheet.

Acknowledgements

The work of the U.S. Ice Drilling Program (IDP) is supported by the National Science Foundation under Cooperative Agreement No. 1836328 to Dartmouth and sub-awards to the University of Wisconsin and the University of New Hampshire. The authors thank anonymous reviewers for their constructive comments.

Author details

Mary R. Albert[1]*, Kristina Slawny[2], Jay Johnson[2], Elliot Moravec[2] and Tanner Kuhl[2]

1 IDP-Dartmouth, Thayer School of Engineering, Dartmouth,
Hanover, New Hampshire, USA

2 IDP-University of Wisconsin, Madison, Wisconsin, USA

*Address all correspondence to: mary.r.albert@dartmouth.edu

References

[1] Souchez R, Lorrain R, Tison J-L. Stable water isotopes and the physical environment. Belgeo. 2002;**2**:133-164. DOI: 10.4000/belgeo.16199

[2] Brook EJ. Ice core methods: Overview. In: Scott AE, editor. Encyclopedia of Quaternary Science. Vol. 2. Oxford: Elsevier; 2013. p. 278. ISBN: 9780444536426

[3] Moore GWK, Alverson K, Holdsworth G. Mount Logan ice core evidence for changes in the Hadley and Walker circulations following the end of the little ice age. In: Diaz HF, Bradley RS, editors. The Hadley Circulation: Present, Past and Future. Advances in Global Change Research, v 21. Dordrecht: Springer; 2004. DOI: 10.1007/978-1-4020-2944-8 14

[4] Fisher DA, Wake C, Kreutz K, Yalcin K, Steig E, Mayewski P, et al. Stable isotope records from Mount Logan, eclipse ice cores and nearby jellybean Lake. Water cycle of the North Pacific over 2000 years and over five vertical Kilometres: Sudden shifts and tropical connections. Géographie Physique et Quaternaire. 2004;**58**(2-3):337-352. DOI: 10.7202/013147ar

[5] Osterberg E, Mayewski PA, Fisher DA, Kreutz KJ, Maasch KA, Sneed SB, et al. Mount Logan ice core record of tropical and solar influences on Aleutian low variability: 500-1998 A.D. Journal of Geophysical Research – Atmospheres. 2014;**119**:11,189-11,204. DOI: 10.1002/2014JD021847

[6] Thompson LG, Davis ME, Mosley-Thompson E, Porter SE, Corrales GV, Shuman CA, et al. The impacts of warming on rapidly retreating high-altitude, low-latitude glaciers and ice core-derived climate records. Global and Planetary Change. 2021;**203**:103538. DOI: 10.1016/j.gloplacha.2021.103538

[7] IPCC, 2021: Climate Change 2021: The Physical Science Basis. Contribution of Working Group I to the Sixth Assessment Report of the Intergovernmental Panel on Climate Change [Masson-Delmotte V, Zhai P, Pirani A, Connors SL, Péan C, Berger S, Caud N, Chen Y, Goldfarb L, Gomis MI, Huang M, Leitzell K, Lonnoy E, Matthews JBR, Maycock TK, Waterfield T, Yelekçi O, Yu R, and Zhou B (eds.)]. Cambridge, United Kingdom and New York, NY, USA: Cambridge University Press; 2021. pp. 33–144. DOI: 10.1017/9781009157896.002

[8] Kelly MA, Lowell TV, Applegate PJ, Smith CA, Phillips FM, Hudson AM. Late glacial fluctuations of Quelccaya ice cap, southeastern Peru. Geology. 2012;**40**:991-994. DOI: 10.1130/G33430.1

[9] Thompson LG, Mosley-Thompson E, Davis ME, Zagorodnov VS, Howat IM, Mikhalenko VN, et al. Annually resolved ice core records of tropical climate variability over the past ~1800 years. Science. 2013;**340**:945-950. DOI: 10.1126/science.1234210

[10] Zagorodnov VS, Thompson LG. Thermla electric ice-core drills: History and new design optins for intermediate-depth drilling. Annals of Glaciology. 2014;**55**(68):322-330. DOI: 10.3189/2014AoG68A012

[11] Clifford HM, Potocki M, Rodda C, Dixon D, Birkel S, Handley M, et al. Prefacing unexplored archives from central Andean surface-to-bedrock ice cores through a multifaceted investigation of regional firn and ice core

glaciochemistry. Journal of Glaciology. 2023;**69**(276):693-707. DOI: 10.1017/jog.2022.91

[12] Yarleque C, Vuille M, Hardy DR, Timm OE, De la Cruz J, Ramos H, et al. Projections of the future disappearance of the Quelccaya ice cap in the Central Andes. Nature. 2018;**18**(8):15564. DOI: 10.1038/s41598-018-33698-z

[13] Bentley CR, Koci B, Augustin LJM, Bolsey RJ, Green JA, Kyne JD, et al. Ice drilling and coring. In: Bar-Cohen Y, Zacny K, editors. Drilling in Extreme Environments. Weinheim, Germany: Wiley-VCH; 2009. ISBN:978-3-527-40852-8

[14] Chandler DM, Hubbard A. Widespread partial-depth hydrofractures in ice sheets driven by supraglacial streams. Nature Geoscience. 2023;**16**:605-611. DOI: 10.1038/s41561-023-01208-0

[15] Chen C, Zhang H, Liu S, Jin C, Chen Y, Zhang N, et al. Hyrdraulic fracturing in ice boreholes: Theory and tests. Polar Science. 2019;**19**:40-48. DOI: 10.1016/j.polar.2018.10.003

[16] Talalay PG. Sublgacial till and bedrock coring. Cold Regions Science and Technology. 2013;**86**:142-166. DOI: 10.1016/j.coldregions.2012.08.009

[17] Goodge JW, Severinghaus JP. Rapid access ice drill: A new tool for exploration of the deep Antarctic ice sheets and sublacial geology. Journal of Glaciology. 2016;**62**(236):1049-1064. DOI: 10.1017/jog.2016.97

[18] Goodge JW, Severinghaus JP, Johnson J, Tosi D, Bay R. Deep ice drilling, bedrock coring and dust logging with the rapid access ice drill (RAID) at Minna Bluff, Antarctica. Annals of Glaciology. 2021;**62**(85-86):324-339. DOI: 10.1017/aog.2021.13

[19] Albert MR, Slawny KR, Boeckmann GV, Gibson CJ, Johnson JA, Makinson RKJ. Recent innovations in drilling in ice. In: Bar-Cohen Y, Zacny K, editors. Advances in Terrestrial Drilling: Ground, Ice and Underwater. 1st ed. Boca Raton, FL: CRC Press; 2021. pp. 157-220. ISBN: 9780367674861

[20] Boeckmann GV, Gibson CJ, Kuhl TW, Moravec E, Johnson JA, Meulemans Z, et al. Adaptation of the Winkie drill for subglacial bedrock sampling. Annals of Glaciology. 2020;**62**(84):109-117. DOI: 10.1017/aog.2020.73

[21] Kuhl T, Gibson C, Johnson J, Boeckmann G, Moravec E, Kristina SK. Agile sub-ice geological (ASIG) drill development and Pirrit Hills field project. Annals of Glaciology. 2021;**62**(84):53-56. DOI: 10.1017/aog.2020.59

[22] Briner JP, Walcott CK, Schaefer JM, Young NE, MacGregor JA, Poinar K, et al. Drill-site selection for cosmogenic-nuclide exposure dating of the bed of the Greenland ice sheet. The Cryosphere. 2022;**16**:3933-3948. DOI: 10.5194/tc-16-3933-2022

[23] Alley RB. Ice core evidence of abrupt climate changes. PNAS. 2000;**97**(4):1331-1334. DOI: 10.1073/pnas.97.4.1331

[24] Lüthi D, Le Floch M, Bereiter B, Blunier T, Barnola JM, Siegenthaler U, et al. High-resolution carbon dioxide concentration record 650,000-800,000 years before present. Nature. 2008;**453**:379-382. DOI: 10.1038/nature06949

[25] Brook E. Windows on the greenhouse. Nature. 2008;**453**:291-292. DOI: 10.1038/453291a

Chapter 3

Perspective Chapter: What Sort of Ice Dynamics Are Crevasse Fill Ridges Connected with? – Research Overview

Anna Orłowska

Abstract

The chapter presents a discussion on crevasse fillings – one of a group of glacial forms whose formation is assigned by various authors to two peripheral dynamics of ice-sheet masses: stagnant ice and surging glaciers. Examples from literature of crevasse filling formation in these two states of dynamics are presented. The author makes an overview of documented in the literature examples of crevasse fillings forming in stagnant and surging ice and discusses the differences between the group formed in stagnant and the group formed in surging ice. In the conclusions, the author arguments assigning crevasse fillings to the surge dynamics of ice masses whereas glacial forms, which develop in stagnant ice crevasses, should not be termed crevasse fillings, but interpreted e.g. as kames, eskers and hummocky moraines.

Keywords: crevasse fill ridges, ice crevasse, stagnant ice, surging ice, glacier/ice sheet

1. Introduction

Among the various forms of glacial relief in glaciated areas, both in the Pleistocene and today, a significant group of them is associated with a specific state of ice dynamics: balanced, positive or negative. The most typical examples of this include: (1) accumulative end moraines, which are an indicator of a balanced ice balance [1, 2]; (2) end moraines characterized by a positive ice balance [1, 3]; (3) kames, which are a record of the negative ice balance [1, 4–6].

However, there are forms of glacial relief whose origin still remains problematic in the context of assigning them to one dynamic state of ice. This group includes crevasse forms, which in the literature are examples of forms attributed to two extreme dynamic states of ice (according to Orłowska [7]): stagnant [8–12] and surging [13–20]. Against this background, controversial issues arise regarding: the term crevasse forms, used for relief forms formed in ice crevasse under conditions of different ice dynamics, and the characteristic features of these forms.

IntechOpen

In the context of the above issues, the following issues are worth considering: (1) a review of examples of forms formed in ice crevasses documented in the literature, (2) a comparison of their features and an attempt to explain their genesis and (3) assigning them to a specific state of glaciers/ice-sheet dynamics.

2. Ice crevasse

The place where crevasse deposits are formed is an ice crevasse. Benn & Evans [21] define this term as a space in the ice created by ice fracture. This occurs when glacial ice does not move fast enough to allow the ice mass to adjust to its shape under the stress. Earlier, Sharp [13] indicated what ice dynamics should be associated with the formation of crevasses, stating that a crack, which is the place where a crevasse develops, can only occur during the active phase of glacier movement.

Similarly, the term "crevasse" was classified by Terpiłowski [5], specifying it as a crack in the ice, conditioned by static and dynamic stresses in the ice. Additionally, he separated the term "ice crevasse" from other, similar concepts, i.e. "ice rift" (defined as a depression in the ice, also reaching the sub-ice surface, resulting from uneven surface ablation) and "ice crevasse" (defined as an elongated, narrow a depression in the ice resulting from the degradation of the roof of an ice tunnel).

The formation of fractures defined in this way, explained in the literature by the distribution of compression and tension stresses in the ice, became the basis for the construction of models for their formation. Nye's [22] best-known model concerned the formation of crevasses in mountain glaciers as a result of shear stress generated by glacier movement. Nye [22] considered three situations of the formation of cracks in the ice as a result of the action of shear stresses (**Figure 1A**): (a) at the interface between the glacier and the valley walls as a result of the friction of the ice on the valley slopes; (b) in the central part of the glacier, where the ice tension flow dominates,

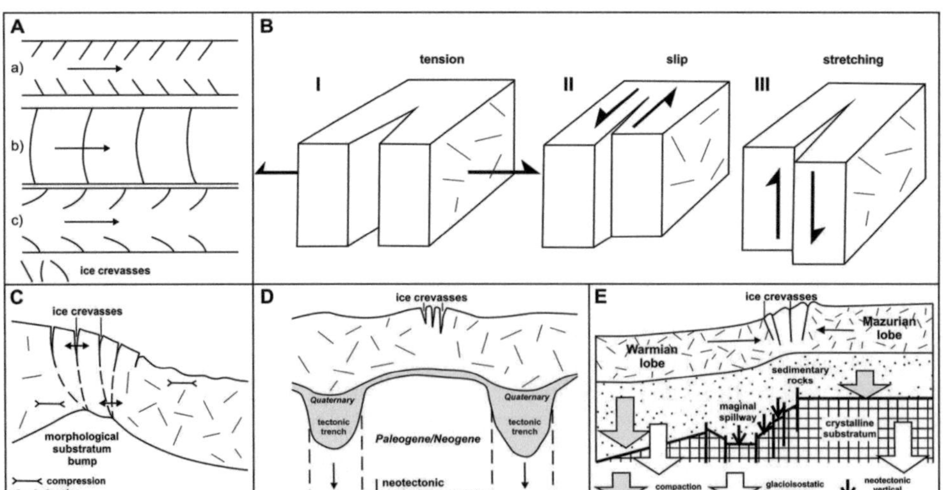

Figure 1.
Models of crevasse formation: (A) according to Nye [22]; (B) according to Benn et al. [23]; (C) according to Röthlisberger & Lang [24]; (D) according to Nitychoruk [25]; (E) according to Morawski [28].

which causes the expansion of fractures oriented transversely to the direction of flow; (c) in the marginal part of the ice, where compressive ice flow occurs.

The latest model by Benn et al. [23] presents three ways of creating cracks as a result of (**Figure 1B**): I – opening mode – occurring as a result of the tension of the walls of cracking ice; II – sliding mode – causing a crack along the shear plane in the same direction as the direction of the shear stress; III – tearing mode – occurring at right angles to the shear direction. Several examples of fracture formation documented in the literature fit these models.

Model I corresponds to a situation in which the ice sheet entered the morphological obstacles present in its base. As a result of the action of tensile stresses, the ice became sealed above the subsurface elevation (e.g., [24]); **Figure 1C**). Such an example was presented by Nitychoruk [25], who suggested subsurface subsidence along Paleogene-Neogene tectonic trenches in the South Podlasie Lowland, activated as a result of ice loading. (**Figure 1D**). Model II corresponds to the most frequently documented cracks at the junction of a glacier/land ice sheet with an ice stream moving much faster than the surrounding ice [26]. Such examples are recorded in modern Antarctic ice sheets [27]. The result of these processes is crevasse forms observed in the post-glacial relief of areas glaciated in the Pleistocene (e.g., [25]). Model III corresponds to the origin of the formation of cracks in the ice located above the tectonic framework, resulting from activating isostatic movements of the ground. Such an example was presented by, among others, Morawski [28, 29], who documented the formation of the interlobe zone, and thus the cracks separating the Vistula glaciation ice sheet into the Warmian and Masurian lobes, thanks to the vertical neotectonic movements of blocks of crystalline basement (**Figure 1E**).

In the fractures defined in this way, numerous deposits of fracture forms, deposited in stagnating and charging ice, have been documented, examples of which are presented in the next chapter.

3. Overview of crevasse forms

3.1 Forms in crevasses of stagnant ice

Crevasse fillings were introduced into the canon of glacigenic forms by Flint [8], based on studies of the glacial relief of the Connecticut area in North America. This author separated them from the group of eskers, pointing out the differences between them. These differences were expressed in: (1) location - crevasse forms were to occur at a certain distance from terminal or recessional moraines or between them, but without a direct connection with them (connection with terminal moraines is a feature often found in eskers, oriented perpendicular to frontal moraine sequences and directly reaching them); (2) morphological features - crevasse forms are short, single and individual embankments that do not show - typical of eskers - traces of connections into long sequences of embankments with irregular ridges and orientation parallel to the direction of ice advance; (3) geological structure - fissure forms are composed of fine-grained sediments, i.e. fine-sandy and silt/clay without clay cover, deposited in a glacial lake environment, i.e. they are different from eskers, composed of coarse-grained sediments deposited in a glacifluvial environment. Flint [8] did not document deformations in the sediments building crevasse forms, which are characteristic of esker deposits covered with basal clay (e.g., [30, 31]).

Crevasse forms with similar morphological features, referring to the work of Flint [8], were documented by Johnson [10], who, while examining the Donjek glacier in the St. Mountains, Elijah, in the Yukon area (Canada), noted a large number of short, single and isolated embankments in cracks parallel to the direction of glacier movement and radiating to the lobe shape of the front. However, the material building these forms was runoff clay. Johnson [10] explained the origin of these forms by the deposition of ablative material, i.e., clay flowing into open crevasses in the glacier; gaps formed during the ice surge, but filled already in the stage of ice stagnation.

Forms formed in the crevasses of stagnant ice were also the object of interest of Polish researchers (including [9, 32–34]). A model of the formation of crevasses as a result of tensions above the ground hump, and then shaping forms in them in the already stagnant Pleistocene ice sheet by filling them with supraglacial sediments: (A) glacifluvial, was presented by Bartkowski [9] in the area of the Wielkopolska Lowland, (B) glacilimnic, presented by Klimek [32] in the area Lesser Poland Upland. However, both authors classified such forms as kames - including Bartkowski [9] as glacifluvial kames. Similar fissure forms near Kornica [33], filling fissures in the stagnant Pleistocene ice sheet, composed of sediments with a fan-delta sequence atypical for fissure forms, and developed above a basement hump (**Figure 2A**), were reinterpreted and classified as glacidelta kames by Godlewska & Terpiłowski [34].

An analogous mechanism of the formation of cracks in the ice stagnating above the ground hump, as well as the formation of forms within them, but with a significantly different internal structure, was presented by Eyles et al. [11]. Based on research on the glaciated area in Canada, they proposed a model of two groups of forms deposited in the so-called cracks, hummocky moraine and linear disintegration

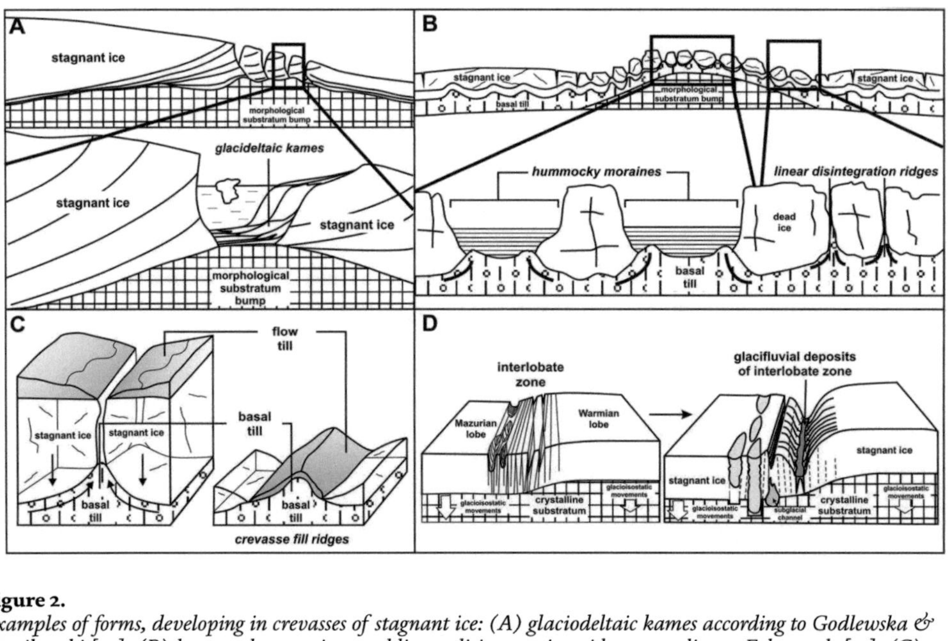

Figure 2.
Examples of forms, developing in crevasses of stagnant ice: (A) glaciodeltaic kames according to Godlewska & Terpiłowski [34]; (B) hummocky moraines and linear disintegration ridges according to Eyles et al. [11]; (C) crevasse fill ridges according to Friello & Hanson [12] and Dreimanis [35]; (D) interlobate eskers according to Gruszka et al. [38].

ridges. According to the above-mentioned authors, both groups of forms in the floor part are composed of soft clay-rich basal till, which, as a result of strong hydration, was plastically pressed into the cracks as a result of being loaded with stagnant ice blocks. They are therefore characterized by the presence of anticlinal extrusion deflections within the clay (**Figure 2B**). Hummocky moraine are forms deposited in wide crevasses. As a result of pressing the clay, the so-called "circular shafts" giving the shape of doughnuts. In the ceiling, these forms are filled and covered with glacilimnic sediments (fine-grained - sandy-mud, silt and clay), deposited in still waters in the ice crevasse. Their characteristic feature is a flat top. These are therefore forms of subglacial-supraglacial origin. However, linear disintegration ridges are formed in narrow cracks and are composed exclusively of subglacial sediments (**Figure 2B**). The lack of supraglacial sediments in their profile is due to the limitation of the upper part of the rift by an ice ceiling, which results in their characteristic morphology, i.e., sharp-edged, "jagged" ridges.

Analogously shaped forms with a similar internal structure, called crevasse fill ridges, were documented by Friello & Hanson [12] in the area of the United States glaciated by the Laurentian ice sheet. In their opinion, the formation of cracks took place in active ice, but their filling took place between blocks of stagnant ice. Like Eyles et al. [11], they documented a subglacial link in the form of basal clay pressed into the cracks from below. However, the above-mentioned authors also proposed an alternative model with a supraglacial link covering the basal clay with runoff clay (**Figure 2C**). Rift forms with such a subglacial-supraglacial succession of layers were also documented by Dreimanis [35] in ice crevasses of the stagnant ice sheet of the Wisconsinan glaciation in Ontario (Canada).

Forms formed in crevasses of stagnant ice have also been documented in interlobe zones *sensu* Punkari [36]. However, the features composed of sediments filling these cracks have been interpreted as kames or eskers.

The formations composed of glacifluvial sediments (sand, sand-gravel and gravel), covered with runoff clay, located in the interlobe zone of the Laurentian ice sheet, were considered to be kames. They were documented by Santos [37] in the Kent area, Ohio (USA). According to this author, sediment deposition took place in crevasses of stagnant ice, the formation of which took place above the subsurface elevation.

However, eskers, formed in the cracks of the interlobe zones of the Pleistocene ice sheet in northern Poland, were documented by Gruszka et al. [38]. The sediments that constitute them were deposited in narrow crevasses, shaped in the ice as a result of vertical movements of the crystalline substrate, activated by the loading of the ice sheet (**Figure 2D**). These eskers form narrow, long embankments, perpendicular to the line of the maximum extent of the Pomeranian phase of the Vistula Glaciation, composed of glacifluvial sediments (sand, gravel, sand and gravel), deposited in high-energy sedimentary environments and runoff clay on the slopes. Deposition of these sediments occurred in stagnant ice (**Figure 2D**).

To sum up, the documented forms formed in the crevasses of stagnant ice are characterized by the following features according to Orłowska [7]: (1) in terms of morphology - they are ridges, plateaus; (2) geologically - composed of both supraglacial (glacifluvial, glacilimnic, ablative, i.e., runoff clay) and subglacial (basal clay) sediments; (3) in terms of location - oriented both perpendicularly and parallel to the glacier front. In the geomorphological classification, they are interpreted as forms of different origins and with different nomenclature, i.e., crevasse forms, kames, eskers, hummocky moraine or linear disintegration ridges.

3.2 Crevasse forms in surging ice

The pioneer in the study of crevasse formations as indicators of surging glaciers was Sharp [13, 14]. In the forelands of the Eyjabakkajökull and Vatnajökull glaciers in Iceland, he documented embankments with steep slopes (70–80° inclination), up to 2–3 m high, up to 2 m wide and with sharp-edged, irregular ridges. These forms were made of basal clay. The author explained the lack of cover with supraglacial sediments by pressing basal sediments through the cracks to the surface of the thin ice in the marginal zone. He associated their formation with the simultaneous creation of cracks and pressing of basal clay into them during the glacier's surge. Only in the phase of calming down of ice activity after the surge was it possible to statically press the clay already present in the cracks and fill the open cracks with basal clay up to the ice surface (**Figure 3**) [13, 14]. After the ice disappeared, forms appeared on the surface in the form of the so-called clay walls (Polish terminology).

Crevasse squeeze ridges (CSRs) associated with glacial advances were also described by Evans & Rea [15, 16], Evans et al. [17, 18] or Waller et al. [20] from the foreland of Icelandic glaciers, as well as Christoffersen et al. [19] on the foreland of the Elisebreen glacier on Svalbard. In each case, these authors confirmed that these forms took the form of low (1–3 m high) embankments, built of basal clay pressed into the cracks from the bottom and were oriented perpendicular to the directions of glacial movement, thus adopting the concept proposed by Sharp [13] their genesis.

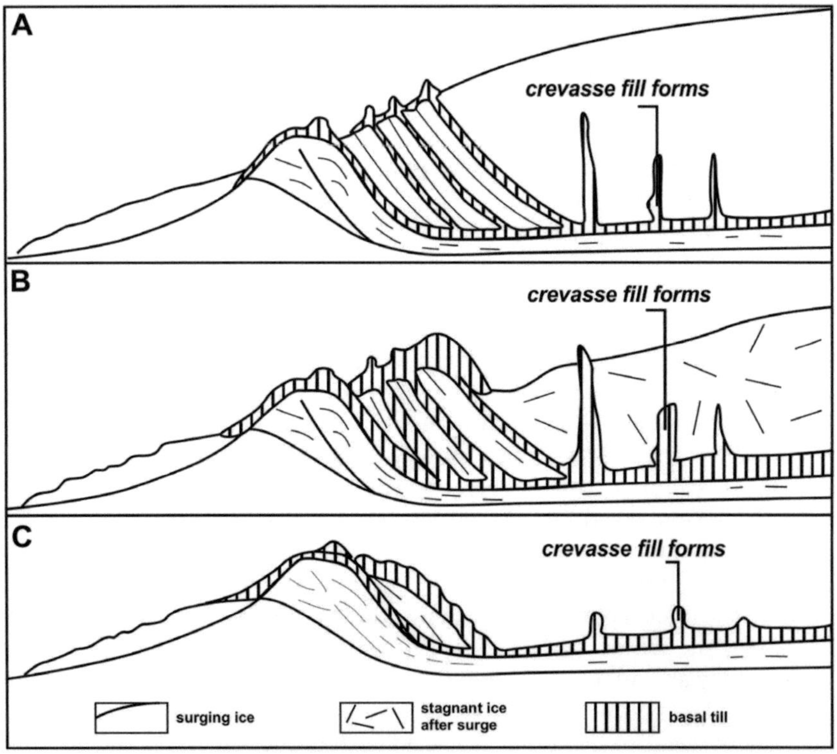

Figure 3.
Examples of crevasse fillings developing in surging ice according to Sharp [13, 14]: (a) situation directly after ice surge; (b) situation at the beginning of the stagnant phase; (c) situation after ice-front recession.

An example of such forms is presented in **Figure 1**. It is worth emphasizing that these are short-lived forms, recorded only in modern glaciers and have no chance of being preserved since the Pleistocene era.

To sum up, the documented forms formed in the cracks of surging ice are characterized by the following features (according to Orłowska [7]): (1) in terms of morphology - they are low, irregular embankments with sharp-edged ridges and steep slopes; (2) geologically - composed exclusively of subglacial sediments, i.e. clayey basal clay pressed into the crack from the ground; (3) in terms of distribution - only parallel to the front of the glacier/ice sheet, and therefore perpendicular to the direction of its advance. In the geomorphological classification, they were assigned various terms: crevasse infills, crevasse fill ridges and CSRs (see, among others, Evans [39] or Benn & Evans [21]).

4. Discussion

Based on the conducted review, there are significant differences in the characteristic features that developed in the crevasses of stagnating and surging ice, i.e. in terms of morphology, geological structure and the distribution of attention at the glacier/ice sheet front (see **Table 1**).

The morphology of the shapes formed in stagnant ice is characteristic of crevasses that widen with differential ablation, hence their final shape in the form of ridges, plateaus with a separate and/or symphonic top. Meanwhile, in surging ice, the top-down closure or only their width results from formations with sharp-edged peaks. Additional, significant definition in the thickness of the sediments of these forms, i.e., up to several dozen meters in stagnant ice and up to several meters in surging ice, two from the above-mentioned wide or maximum opening or closing of the rift, but also from the boundary ablation of the most often thick ice in Pleistocene ice sheets, than the limited limits of marginal mountain glaciers surging.

The geological structure of the forms in the crevasses of stagnant ice, i.e., the glacifluvial, glacilimnic and ablative sediments documented there, are typical of the sediments of the supraglacial subenvironment, flowing into the crevasses from the ice surface, more or less sorted depending on the share of meltwater. The basal clay present in the cracks of stagnant ice is the result of being pressed into the cracks from below as a result of the load of long-lasting stagnant ice. This is a much different process from pressing basal clay into surging ice. In it, when the clay at the foot of the ice sheet/glacier is saturated with water, it enhances the sliding process, reducing friction and facilitating the rapid movement of ice, and thus its tension and the formation of cracks. At the same time, it is dynamically, subglacially pressed into the opening spaces in the ice from below (according to Sharp [13, 14]) and it is the only sediment filling the crevasses of the surging ice (see **Table 1**).

Significant differences in the forms in which sediments are deposited in ice crevasses with extremely different dynamic states also concern the distribution of these forms (see **Table 1**). The involvement of sediments in forms arranged perpendicular or parallel to the ice front or the direction of its advance is related to the distribution of fractures during the transgression of glaciers/land ice sheets. Active glaciers with a normal advance rate are characterized by a distribution of crevasses parallel to the direction of transgression, i.e., perpendicular to the ice front (cf. Evans & Twigg [40]). Therefore, most of the forms deposited in the crevasses, after reaching the maximum extent, of the already stagnant Pleistocene glaciers/land ice sheets, have a

Author	Terminology	Morphology	Lithology	Orientation to ice direction
Crevasse-fill forms in stagnant ice				
Flint [8]	crevasse fillings	short, individual ridges (up to 500 m in length)	glaciolimnic deposits	parallel
Johnson [10]	crevasse fillings	short, individual ridges	glacial deposits (flow till)	parallel
Bartkowski [9]	glaciofluvial kames	short, individual ridges	glaciofluvial deposits	transverse
Klimek [32]	glaciolacustrine kames	flat-topped hillocks	glaciolimnic deposits	lack of information
Dreimanis [35]	crevasse fillings	narrow ridges, elongated hummocks (up to 1–8 m in height, 100–200 m in width, 0.2–1.5 km in length), located above an elongated ridge 5 km in length and 10 m in height	glaciofluvial deposits, glacial deposits (basal till)	parallel
Eyles et al. [11]	hummocky moraine	flat-topped hillocks and hummocks with a doughnut shape (up to 25 m in height)	glaciolimnic deposits, glacial deposits (basal till)	parallel
Eyles et al. [11]	linear disintegration ridges	narrow ridges with acute edges	glacial deposits (basal till)	parallel
Godlewska and Terpiłowski [34]	glaciodeltaic kames	short, individual ridges (up to 1 km in length, 200 m in width, 10 m in height)	glaciodeltaic deposits	transverse
Gruszka et al. [38]	interlobal eskers	ridges (approx. 0.4–1.3 km in length, 150 m in width, approx. 10–15, max. 40 m in height)	glaciofluvial deposits, glacial deposits (flow till)	parallel
Santos [37]	interlobal kames	Belts of asymmetric, round-shaped hummocks with gentle slopes (15–20 m of height)	glaciofluvial deposits, glacial deposits (flow till)	parallel
Crevasse-fill forms in surging ice				
Sharp [13, 14]	crevasse fillings	sharp-crested, narrow ridges (with 2–3 m of height, up to 2 m in width) with steep slopes	glacial deposits (basal till)	transverse
Evans & Rea [15], Evans et al. [18], Christoffersen et al. [19]	crevasse squeeze ridges	ridges (up to 1–3 m in height)	glacial deposits (basal till)	transverse
Evans et al. (2016)	crevasse squeeze ridge corridors	a 200-km-long and 10-km-wide linear assemblage of ridges	glacial deposits (basal till)	transverse

Table 1.
List of forms in crevasses of stagnant and surging ice documented in the references (according to Orłowska [7]).

DOI: http://dx.doi.org/10.5772/intechopen.1005192

direction parallel to the direction of their advance (see **Table 1**). Sometimes, examples of these forms with a transverse orientation are related not to the dynamics of the glacier/land ice sheet itself, but to external conditions causing the ice to seal, e.g. with specific ground conditions, i.e., its morphology (see, among others, Terpiłowski [4]; Godlewska & Terpiłowski [34]), also resulting from isostatic movements of this substrate (e.g., Nitychoruk [25]; Gruszka et al. [38]). Meanwhile, the distribution of fractures in the charging ice is oriented only transversely to the direction of ice advance, because during the surge the glacial tongue is stretched longitudinally and separated into separate blocks as a result of the action of tension stresses. They favour the formation of crevasses parallel to the ice front as a result of the rapid pace of ice movement. Therefore, as a result, formations are created in crevasses located only parallel to the front of the charging glaciers (see **Table 1**).

The above differences in morphology, geological structure and the distribution of forms in the crevasses of stagnating and charging ice may also result from the chronology/sequence of the formation of the cracks and their filling. The formation of forms in the crevasses of stagnant ice does not occur synchronously with the formation of these crevasses. These are formed when the ice is active, and only then, after the glacier/land ice activity ceases, they are filled. Such paleogeographic conditions lead to the formation of forms that can generally be classified as ice disintegration forms, e.g., kames, hummocky moraines or linear disintegration ridges, but also eskers (**Figure 4**). Meanwhile, in charging ice, the formation and filling of cracks occur synchronously, i.e., the crevasse opening from the bottom is immediately filled with the intruding basal clay and pressed within it during the surge, as well as after it stops (**Figure 4**). Such paleogeographic conditions are possible only for surging glaciers, which is why such forms - according to the examples presented in the literature - are considered in the world literature to be crevasse forms in the strict sense (**Figure 4**) and are associated with the landscape of surging glaciers (*terrestrial surging glacier landsystem* as defined by Evans & Rea [16]).

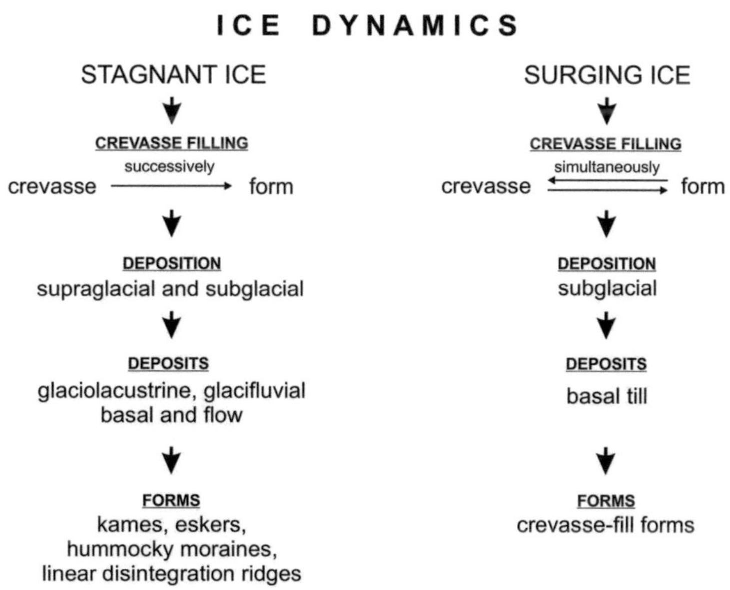

Figure 4.
Ice dynamics and forms deposited in crevasses of stagnant and surging ice (according to Orłowska [7]).

5. Conclusion

Filling ice crevasses, in the context of their genesis, has so far been associated with extreme dynamic states of ice: stagnant or charging. A review of the literature documenting slotted forms allows us to draw the following conclusions:

1. Forms filling the crevasses of stagnant ice are most often embankments, plateaus, made of glacifluvial, glacilimnic and glacial sediments, with various arrangements: parallel and perpendicular to the direction of ice advance. These forms are characteristic of ice disintegration. The places of their deposition, i.e., crevasses, were formed during active ice, and the deposition in the crevasses took place in the final phase of activity and in the initial phase of ice stagnation.

2. Meanwhile, crevasses forms in the strict sense correspond to glacial forms in the form of low, sharp-edged embankments, the deposition of which in the crevasses occurred synchronously with the formation of ice crevasses parallel to the front and perpendicular to the direction of ice advance. Such conditions of their formation are only possible in surging glaciers. The deposition of such forms takes place from the bottom up into the crevasses by pressing basal clay into them.

Examples of forms documented in the literature show that crevasses forms defined in this way only occur in modern glaciated areas of high latitudes.

Author details

Anna Orłowska
Department of Geomorphology and Palaeogeography, Institute of Earth and Environmental Sciences, Maria Curie-Skłodowska University, Lublin, Poland

*Address all correspondence to: anna.orlowska@mail.umcs.pl

References

[1] Flint RF. Glacial and Quaternary Geology. New York: Wiley; 1971

[2] Zieliński T. Moreny Polski północno-wschodniej – osady i warunki sedymentacji. Vol. 1325. Katowice: Prace Naukowe; 1992

[3] Bennett MR. The morphology, structural evolution and significance of push moraines. Earth Science Reviews. 2001;**53**(3-4):197-236

[4] Terpiłowski S. Genesis of transverse kame trains in eastern Poland. Sedimentary Geology. 2007;**193**:59-69

[5] Terpiłowski S. Kemy jako wskaźnik deglacjacji Niziny Podlaskiej podczas zlodowacenia Warty. Lublin: Wyd, UMCS; 2008

[6] Livingstone SJ, Evans DJA, Cofaigh ÓC, Hopkins J. The Brampton kame belt and Pennine escarpment meltwater channel system (Cumbria, UK): Morphology, sedimentology and formation. Proceedings of Geological Association. 2010;**121**:423-443

[7] Orłowska A. Crevasse-fill forms – Bridging the gap in glacial geomorphology between east and west based on a case study from eastern Poland. Quaternary International. 2022;**617**:59-72

[8] Flint RF. Eskers and crevasse fillings. American Journal of Sciences. 1928;**15**:410-416

[9] Bartkowski T. O formach strefy marginalnej na Nizinie Wielkopolskiej. Prace Komisji Geograficzno-Geologicznej. 1967;**7**(1):260

[10] Johnson PG. Recent crevasse fillings at the terminus of the Donjek glacier, St. Elias Mts., Yukon territory. Quaestiones Geographicae. 1975;**2**:53-59

[11] Eyles N, Boyce JI, Barendregt RW. Hummocky moraine: Sedimentary record of stagnant Laurentide ice sheet lobes resting on soft beds. Sedimentary Geology. 1999;**123**:163-174

[12] Friello C, Hanson G, UnpublishedAnatomy of a Crevasse Fill Ridge in Clara's Woods. New York: Stony Brook; 2008

[13] Sharp M. Crevasse fill ridges – A landform type characteristic of surging glaciers? Geografiska Annaler. 1985;**67A**(3-4):213-220

[14] Sharp M. Surging glaciers: Geomorphic effects. Progress in Physical Geography. 1988;**12**:533-559

[15] Evans DJA, Rea BR. Geomorphology and sedimentology of surging glaciers: A landsystems approach. Annals of Glaciology. 1999;**28**:75-82

[16] Evans DJA, Rea BR. Surging glacier landsystem. [W]. In: Evans DJA, editor. Glacial Landsystems. London: Edward Arnold; 2005. pp. 259-288

[17] Evans DJA, Lemmen DS, Rea BR. Glacial landsystems of the southwest Laurentide ice sheet: Modern Icelandic analogues. Journal of Quaternary Sciences. 1999;**14**(7):673-691

[18] Evans DJA, Twigg DR, Rea BR, Shand M. Surficial geology and geomorphology of the Bruarjokull surging glacier landsystem. Journal of Maps. 2007;**3**:349-367

[19] Christoffersen P, Piotrowski JA, Larsen NK. Basal processes beneath

an Arctic glacier and their geomorphic imprint after a surge, Elisebreen, Svalbard. Quaternary Research. 2005;**64**:125-137

[20] Waller RI, Van Dijk T, Knudson O. Subglacial bedforms and conditions associated with the 1991 surge of Skeidararjökull, Iceland. Boreas. 2008;**37**:179-194

[21] Benn D, Evans DJA. Glaciers and Glaciation. London: Edward Arnold; 1998

[22] Nye JF. The mechanics of glacier flow. Journal of Glaciology. 1952;**2**(12):82-93

[23] Benn DI, Warren CR, Mottram RH. Calving processes and dynamics of calving glaciers. Earth Science Reviews. 2007;**82**:143-179

[24] Röthlisberger H, Lang H. Glacial hydrology. [W]. In: Gurnell AM, Clark MJ, editors. Glaciofluvial Sediment Transfer. An Alpine Perspective. London: Wiley & Sons, Ltd.; 1987. pp. 207-284

[25] Nitychoruk J. Stratygrafia plejstocenu i paleogeomorfologia południowego Podlasia. Międzyrzec Podlaski: Wyd. Towarzystwa Przyjaciół Międzyrzeca Podlaskiego; 1995

[26] Swithinbank CWM. Ice streams. Polar Record. 1954;7:185-186

[27] Bentley CR. Antarctic ice streams: A review. Journal of Geophysical Research. 1987;**92**:8843-8858

[28] Morawski W. Neotectonics induced by ice-sheet advances in NE Poland. Geologos. 2009;**15**:199-217

[29] Morawski W. Differences in the regional stratigraphy of NE Poland caused by vertical movements due to glacioisostasy. Geologos. 2009;**15**:235-250

[30] Krupa A. Depozycja i postdepozycyjne deformacje osadów ozu – stanowisko Wielowicz-Wielowiczek, Pojezierze Krajeńskie. [W:] M. Pisarska-Jamroży, Z. Babiński (red.), Plejstoceńskie środowiska sedymentacyjne Pojezierza Pomorskiego. Wyd. Uniwersytetu Kazimierza Wielkiego, Bydgoszcz; 2009. pp. 84-89

[31] Gruszka B, Dobracka E, Langner A. Poligenetyczny oz w Kluczkowie koło Świdwina, NW Polska. [W]. In: Pisarska-Jamroży M, Dobracki R, editors. Osady glacigeniczne fazy pomorskiej od Uckermark po Równinę Białogardzką. Nauk., Poznań: Bogucki Wyd; 2011. pp. 85-96

[32] Klimek K. Wpływ rzeźby podłoża lądolodu na wykształcenie i budowę form kemowych w NW części Wyżyny Małopolskiej. Folia Quaternaria. 1969;**30**:13-16

[33] Albrycht A. Detailed Geological Map of Poland at 1:50 000 Scale, Sarnaki Sheet. Warszawa: PIG; 2004, [in Polish]

[34] Godlewska A, Terpiłowski S. Transverse, supraglacially-derived crevasse infillings in Pleistocene ice-sheet: Genesis and sedimentary record. Geomorphology. 2012;**161-162**:73-81

[35] Dreimanis A. Landforms and structures of the waterlain west end of St. Thomas moraine, SW Ontario, Canada. Geomorphology. 1995;**14**:185-196

[36] Punkari M. Glacial and glaciofluvial deposits in interlobate areas of the Scandinavian ice sheet. Quaternary Science Reviews. 1997;**16**:741-753

[37] Santos JB. Late Wisconsinan glacial geomorphology of the Kent Interlobate Comples, Ohio, USA. Finisterra. 2012;**47**(93):65-84

[38] Gruszka B, Morawski W, Zieliński T. Sedimentary record of a Pleistocene ice-sheet interlobate zone (NE Poland). Geologos. 2012;**2**:65-81

[39] Evans DJA. Glacial Landsystems. London: Edward Arnold; 2005

[40] Evans DJA, Twigg DR. The active temperate glacial landsystem: A model based on Breiðamerkurjökull and Fjallsjökull, Iceland. Quaternary Science Reviews. 2002;**21**:2143-2177

Section 3

Climate Change during an Interglacial

Chapter 4

Spatial Distribution and Shrinkage of Glaciers since the 1990s in the Transboundary Kailash Sacred Landscape

Sudan Bikash Maharjan, Tenzing Choyal Sherpa, Finu Shrestha and Binaya Pasakhala

Abstract

Glaciers are an inherent component of the landscape, culture, and environment in the high mountain areas of the Himalayas. Glacier distribution and its impacts are diverse in various landforms and landscape scales. Past studies on glaciers have focused on the individual, river basins, or at the country level, but none have been analyzed at the transboundary landscape level. This study investigates the spatial distribution of glaciers in the transboundary Kailash sacred landscape (KSL) at multiple time scales. The study revealed that 3.8% of the total landscape area is covered by glaciers. Altogether, 1941 glaciers were identified, covering 1169.04 ± 27.71 km^2 area in 2020, most of which are highly concentrated in the southern part of the landscape in the Kali and West Seti basins. From 1990 to 2020, these glaciers have retreated significantly by 25.5% of their area. Oppositely, due to the shrinkage and fragmentation of individual glaciers, the number of glaciers has increased by 7.8%. The glacier retreat rate is higher at elevations below 5500 masl, and glaciers below 3500 masl have disappeared completely after 2010. Systematic long-term glacier monitoring is required, and the data can be utilized to project water resource availability based on various climatic and glacio-hydrological models.

Keywords: mountain glaciers, glacier change, debris-covered glaciers, remote sensing, transboundary landscape

1. Introduction

Glaciers are an inherent component of the landscape in the high mountain areas of the Hindu Kush Himalaya (HKH) region, having religious and cultural significance in many societies [1, 2]. They play a pivotal role in freshwater supply for agriculture, industrial and domestic use, hydropower generation, and recreation [3, 4]. The HKH region, often dubbed as the third pole, holds 60,000 km^2 of glacier area [5], equivalent to 8.5% of the world's total. Global glacier data is available from various initiatives such as the Global Land Ice Measurements from Space (GLIMS) [6],

Randolph Glacier Inventory (RGI) [7], the Gamdam Glacier Inventory (GCI) [8], and the Second Chinese Glacier Inventory (CGI) [9]. These databases provide an understanding of the regional coverage of glaciers for a particular period and are widely used in cryo-hydrological modeling [10]. The databases were compiled and generated using various mapping methods and data sources, and their accuracy varies across subregions and basins. For the systematic change analysis, it is necessary to generate the data from consistent data sources and methods.

Glaciers are very sensitive to changes in temperature. They have been retreating, shrinking, and thinning rapidly [11–18] due to a rise in global temperature, most notably since the late 1970s [19]. The trend of temperature rise is projected to continue resulting in the expected loss of approximately 36% of glacier ice by the end of this century in the HKH region [18, 20]. Many studies show the decrease in glacier area throughout the Himalayas and Tibetan Plateau [12, 13, 15, 21–23]. The changes vary spatially depending on size, shape, morphological and topographical characteristics, and local and regional climatic conditions. The studies at the country level show a drastic retreat of glaciers – in Nepal, it has retreated by 24% and in Bhutan by 23% during the 1980–2010 period, and in Afghanistan by 13.8% between 1990 and 2015 [12, 13, 24, 25].

Alongside retreat and shrinkage, glaciers in the region are also thinning significantly and losing ice mass. Between 2000 and 2016, the loss of ice mass across the Himalayas has doubled compared to 1975–2000 [26]. However, in some areas, especially in the Karakorum (northern Pakistan), glaciers are advancing mostly in individual surges, without significant changes in glacier area [18, 27, 28].

The recession of glaciers in the region also resulted in an increase in debris-covered ice [29, 30], and the elevation of debris-covered parts is moving upward [31]. Thick debris cover on ice insulates the ice from melting, but if it is covered by a thin layer, the melting rate will further accelerate. In addition, the formation of new glacial lakes and expansion of glacial lakes either attached to or close by glaciers have also increased. This increases the risks of glacial lake outburst flood (GLOF), threatening the life and infrastructure of communities living downstream [32, 33].

Past studies on glaciers have focused on individual, river basins or at the country level, but none have been analyzed at the transboundary landscape level. Glaciers in the HKH region are water sources for transboundary rivers such as the Ganges, Brahmaputra, and Indus. Therefore, the shrinking of glaciers and the impact on river hydrology will have serious implications for the population living downstream. Hence, the information on individual scale or river basins or mountain ranges is not sufficient to cover the various issues and impacts at the landscape level. This research aims to understand the status of glaciers in a transboundary landscape and to analyze their physical characteristics and changes in the area between 1990 and 2020. The information from this study will support effective conservation and management of the landscape.

2. Study area

The Kailash Sacred Landscape (KSL) is a transboundary landscape extending across the southwestern portion of the Tibet Autonomous Region (TAR) of China, four districts in the western region of Nepal, and the northeastern flank of Uttarakhand State in India [34]. The KSL covers an area of 31,000 km^2 and has three major rivers flowing through it – Mahakali (Kali), west Seti, and Humla-Karnali; tributaries of Ghaghara River and one major Manasarovar interior basin (**Figure 1**). The Ghaghara River, a major left-bank tributary of the Ganges, originates as the Karnali

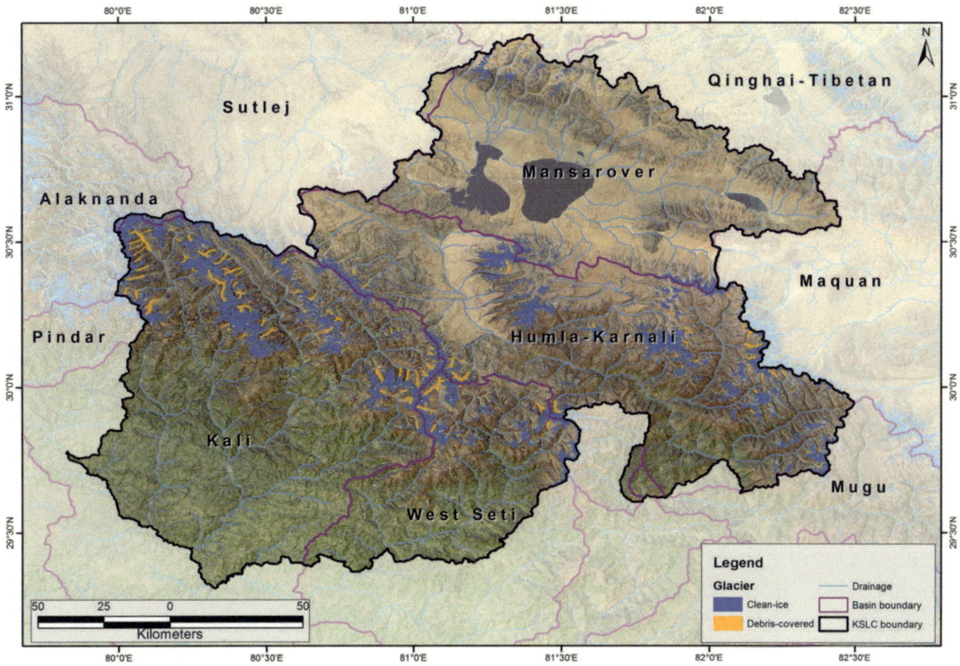

Figure 1.
Distribution of glaciers in various river basins in the Kailash sacred landscape.

River (known as Kongque He in Chinese) in the higher Himalayas of the southern Tibetan Autonomous Region (TAR), China. Following southeast through Nepal and crossing the Siwalik range, it splits into branches and rejoins south of the Indian border. The river then continues its course through Uttar Pradesh and Bihar states, eventually mearing with Ganges. Almost 75% of the KSL is covered by the tributaries of the Ganga River, and around 25% of the area is covered by the Manasarovar inland catchment. Less than 1% of the area falls in the Indus Basin.

The KSL represents a diverse and multicultural landscape [35]. The landscape is characterized by numerous sacred sites, including high-altitude lakes, snow peaks, glaciers, and a network of religious places across the three countries [35]. The Mt. Kailash and Manasarovar Lake are the most important holy sites for five religions and a destination for many pilgrims. It shows great variations in altitude and topography. The elevation ranges from 380 masl, the lowest along the river valley of Kali, to 7500 masl, the highest peak in the Humla basin in the northern part of the landscape. The ecosystems of the landscape vary widely from subtropical to temperate, alpine, and cold high-altitude desert types. The melting of snow, glaciers, and permafrost in the high mountain areas of the landscape are major sources of freshwater for all the major rivers and tributaries, especially during the dry season.

3. Data and methods

3.1 Data sources

The study used various series of Landsat (Thematic Mapper (TM), Enhanced Thematic Mapper (ETM) + and OLI) images as it has a long historical record and

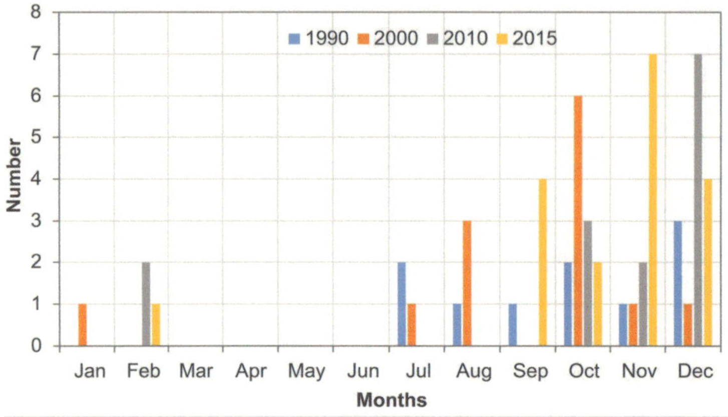

Figure 2.
Seasonal distribution images used for glacier mapping with respect to representative years.

is freely accessible through the USGS web portal. The acquired images for all time periods (1990, 2000, 2010, and 2020) were from October to December, except for a few images from January to September (**Figure 2**). One-year buffer time was used to get the best quality images, except in 1990, for which a 2-year buffer was used, if the images of the selected period were of low quality. Data gaps due to scan line error in ETM+ images, especially for 2010, were anticipated by overlaying the two images of the subsequent month as the seasonal changes within 2 to 3 months are nominal. Overall, seven tiles of Landsat images cover the complete area of KSL in which a major part of the glaciated area is covered by two tiles (144/39 and 145/39). Altogether, 40 Landsat images were used to prepare four-time data from 1990 to 2020 and included 10 Thematic Mapper (TM) images for 1990, one Thematic Mapper (TM) and 12 Enhanced Thematic Mapper plus (ETM+) images for 2000, 8 TM and 6 ETM+ images for 2010, 4 OLI images for 2020 glacier data.

Terrain characteristics such as slope, elevation, and aspect are also important parameter for the identification of glaciers. Higher resolution DEMs such as ASTER GDEM and SRTM 30 m are available, but they consist of many voids in the higher Himalayan region. Hence, this study used the void-filled SRTM 90 m DEM for glacier mapping and data generation.

3.2 Glacier mapping method

The present study adopted a semi-automatic object-based image classification method [12, 13, 36], as illustrated in **Figure 3**. Primary Landsat images of 2020 were used to create image objects based on homogenous characteristics of pixels. The multi-resolution segmentation process was applied with an appropriate scale factor to create image objects so that the boundary of image objects exactly follows the glacier boundary. These image objects were classified by developing separate rule sets for clean-ice (CI) and debris-covered (DC) glaciers based on the spectral and spatial characteristics of the image objects. We applied widely used spectral indices NDSI for the classification of CI objects as the primary rule, and other various indices such as NDVI, LWM, and other spectral and spatial characteristics, including terrain information and area, were used for filtering misclassified objects from the primary rule. After the classification of CI glacier objects, a similar process was applied for

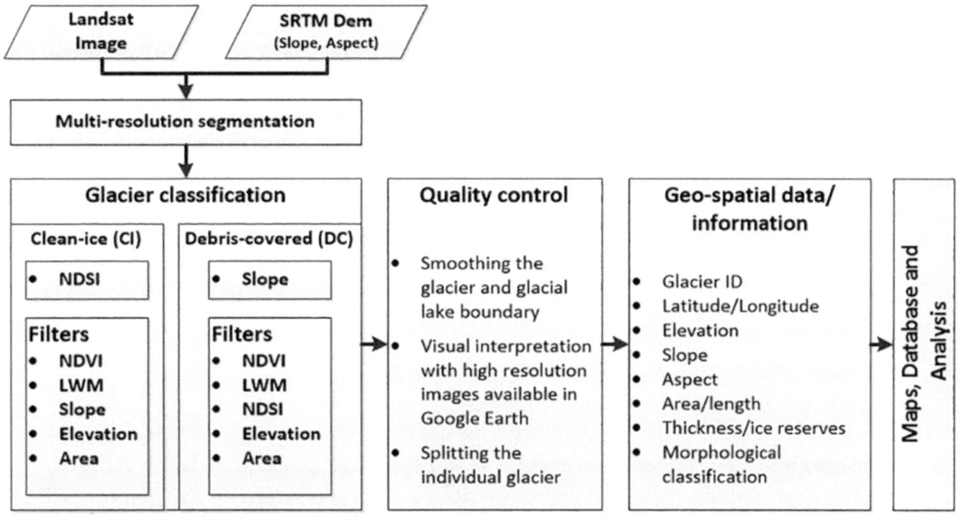

Figure 3.
Flow chart of glacier mapping and glacier database generation (modified [36], p. 12).

classifying DC glaciers using slope as the primary classification rule in the remaining unclassified objects. All the final classified CI and DC image objects were merged and then exported to a vector file.

The vector-based glacier polygons were thoroughly checked by overlaying on the respective images, and some manual correction was done in shadow, seasonal snow cover, turbid/frozen proglacial lakes, and especially the boundary of the DC glacier and its terminus. The glacier boundary was smoothed automatically to remove the zigzag boundary formed due to the pixel size of the images. Each glacier within the glacier polygons was manually separated based on the hydrological flow direction [36–38]. The detailed characteristics of individual glaciers were calculated, and analysis was done in the ArcGIS environment using DEM and various tools [36–38].

The glaciers of other time periods (2010, 2000, and 1990) were prepared manually by using the same final glacier data of 2020 overlaying on the respective images of 2010, 2000, and 1990 so that there are no discrepancies of overlapping in all the time series datasets.

3.3 Uncertainty

The uncertainty in glacier boundaries typically depended on the quality of images used, interpretation made by individuals, and contrast between the glacier and adjacent features [39, 40, 13]. The study used little or no snow cover and cloud-covered images for mapping as far as possible to generate quality boundaries from automatic processes. The manual correction was made for overshadows, seasonal snow covers, and low contrast between DC glacier boundaries and adjacent terrain by overlying the respective images and visual cross-checking on high-resolution images available on Google Earth images to maintain the quality of the glacier boundary. In order to minimize discrepancies between decadal time data, the semi-automatically mapped glacier boundaries of 2020 were overlaid on the representative images of 1990, 2000, and 2010 to manually prepare glacier boundaries of the respective time periods. This process reduced most of the errors in the glacier outlines; uncertainty only remains on the position of glacier outlines with respect to the image resolution, which is dependent on and could not be greater than half of its resolution. Hence, the uncertainty of glacier boundaries was calculated as the root mean square of each glacier area from the glacier polygon and the area was calculated as the total number of pixels bounded by the glacier boundary with a buffer (half of the image resolution) increment [13]. Overall, the uncertainty of the glacier area of each time data ranged from 2 to 3% of the total area, which also matched other previous estimates [13, 41, 42].

4. Results

4.1 Overall distribution and shrinkage of glaciers

The study revealed 1941 glaciers covering a surface area of 1169.04 ± 27.71 km^2, which is equivalent to 3.8% of the total landscape area in 2020 (**Table 1**). Almost 13.3% of the total glacier area is covered by debris. The majority of the glaciers (91%) are in the Ghaghara basin, which is one of the major tributaries of the Ganges River. Within the basins, the Kali sub-basin consists of almost 39% of glaciers in KSL, covering 45% of the total area of glaciers in the landscape. Similarly, the second highest number (more than 38%) and area (38%) coverage is in the Humla-Karnali sub-basin. The Manasarovar inland basin only consists of 9% of glaciers, with an area coverage of 6% of the total glacier area in the KSL. Overall, the number of glaciers has increased by 7.8%, while the glacier area has decreased by 25.5% in three decades from 1990 to 2020 (**Figure 4**), with a drastic decrease in recent decades, i.e., 2010–2020 (11.9%), compared to previous decades.

4.2 Distribution and changes in clean-ice(CI) and debris-covered (DC) glaciers

Glaciers are classified into clean-ice (CI) and debris-covered (DC) depending upon their composition. Clean-ice glaciers are formed due to accumulation and compaction of seasonal snow cover, whereas DC glaciers are part of glaciers derived from erosion and transportation of debris and generally overlay the clean-ice portion of glaciers. Out of 1941 glaciers, 112 glaciers have both components of CI and DC, and the rest have only CI. The area covered by DC glaciers is more than 13% of the total glacier area in the KSL. The highest area coverage by DC glacier is in the Kali sub-basin, and the lowest is in the West Seti sub-basin. The Manasarovar inland basin consists only of CI glaciers.

The area percentage of DC glaciers has increased from 8.7 to 13.3%, while CI glaciers decreased from 91 to 87% from 1990 to 2020 (**Figure 5**), with a marked variation in the individual basins. Within this time period, the DC glacier area increased from 6.6 to 9.7% in the Humla-Karnali basin, 11.8 to 17.8% in the Kali basin, and 8.8 to 14.4% in the West Seti basin. Not much change was noticeable in the DC glacier area between 2000 and 2010. The increase in DC and decrease in CI glacier area indicate that glaciers are melting, and exposure of the debris transported along the CI glacier and covering the clean-ice portion. Debris cover on the glacier plays an important role in insulating the glacier from rapid melting [43–45]. In contrast, a thin layer of debris and dust particles on the surface of glacier ice accelerates the melting of the glaciers.

The CI glaciers maintain a higher slope than DC glaciers. In the KSL, the mean slope of glaciers ranges from 10 to 60 degrees, with a negligible number of glaciers less than 10 degrees and higher than 60 degrees. The mean slope of CI glaciers ranged

DOI: http://dx.doi.org/10.5772/intechopen.1004519

Year	Basin	Sub-basin	Glacier Number			Area (km^2)			
			CI	DC	Total	CI	DC	Total	Largest
2020	Ghaghara	Kali	751	57	751	429.94 ± 10.89	93.44 ± 0.9	523.38 ± 12.84	49.46 ± 1.29
		Humla	739	42	739	406.37 ± 8.75	43.81 ± 0.45	450.18 ± 9.49	12.53 ± 0.27
		West Seti	282	13	282	105.81 ± 2.18	17.78 ± 0.1	123.59 ± 2.5	12.33 ± 0.26
		Total	1772	112	1772	942.12 ± 21.75	155.03 ± 1.47	1097.15 ± 24.69	49.46 ± 1.29
	Manasarovar		169	0	169	71.89 ± 2.47	0 ± 0	71.89 ± 2.47	7.13 ± 0.26
	Total		1941	112	1941	1014.01 ± 24.61	155.03 ± 1.47	1169.04 ± 27.71	49.46 ± 1.29
2010	Ghaghara	Kali	738	56	738	503.21 ± 9.16	90.03 ± 0.46	593.24 ± 10.45	50.83 ± 0.9
		Humla	706	43	706	462.16 ± 8.68	46.19 ± 0.3	508.35 ± 9.31	14.30 ± 0.27
		West Seti	266	13	266	126.23 ± 2.12	16.4 ± 0.11	142.63 ± 2.35	13.09 ± 0.22
		Total	1710	112	1710	1091.6 ± 19.91	152.62 ± 0.89	1244.22 ± 22.06	50.83 ± 0.9
	Manasarovar		170	0	170	83.17 ± 2.03	0 ± 0	83.17 ± 2.03	7.58 ± 0.18
	Total		1880	112	1880	1174.77 ± 22.18	152.62 ± 0.89	1327.39 ± 24.42	50.83 ± 0.9
2000	Ghaghara	Kali	735	56	735	542.64 ± 11.17	91.08 ± 0.47	633.72 ± 12.6	52.14 ± 1.03
		Humla	683	42	683	509.47 ± 7.48	46.1 ± 0.33	555.57 ± 7.97	14.52 ± 0.21
		West Seti	255	13	255	138.69 ± 1.94	19.86 ± 0.11	158.55 ± 2.18	14.75 ± 0.21
		Total	1673	111	1673	1190.8 ± 20.76	157.04 ± 0.96	1347.84 ± 22.85	52.14 ± 1.03
	Manasarovar		175	0	175	90.47 ± 2.2	0 ± 0	90.47 ± 2.2	9.37 ± 0.23
	Total		1848	111	1848	1281.27 ± 23.32	157.04 ± 0.96	1438.31 ± 25.51	52.14 ± 0.87
1990	Ghaghara	Kali	709	55	709	608.16 ± 9.7	81.17 ± 0.55	689.33 ± 10.66	54.5 ± 0.22
		Humla	673	41	673	556.07 ± 7.69	39.49 ± 0.31	595.56 ± 8.07	15.5 ± 0.22
		West Seti	243	13	243	168.88 ± 2.11	16.33 ± 0.14	185.21 ± 2.28	16.0 ± 0.2
		Total	1625	109	1625	1333.11 ± 19.47	136.99 ± 1.02	1470.1 ± 20.97	54.5 ± 0.87
	Manasarovar		175	0	175	98.26 ± 1.74	0 ± 0	98.26 ± 1.74	9.7 ± 0.17
	Total		1800	109	1800	1431.37 ± 21.38	136.99 ± 1.02	1568.36 ± 22.91	54.5 ± 0.87

Table 1.
Number and area of clean-ice and debris-covered glaciers in the Kailash sacred landscape.

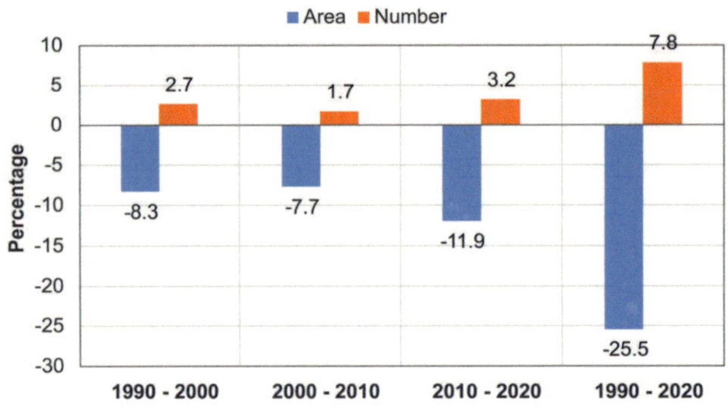

Figure 4.
Percentage changes in number and area of glaciers from 1990 to 2020.

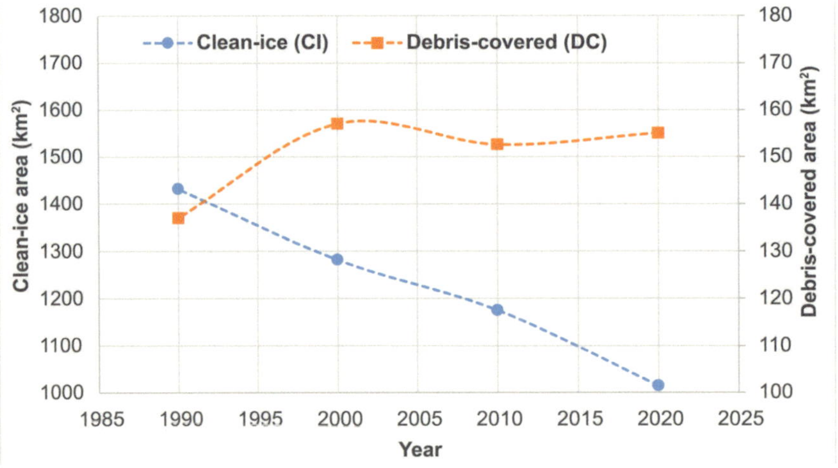

Figure 5.
Changes in DC and CI glacier area from 1990 to 2020 in KSL.

from 23 to 31 degrees, and DC glaciers from 12 to 13 degrees. The mean slope of CI glaciers in the Kali and West Seti basins was slightly higher and ranged up to 31 degrees, whereas the mean slope of Manasarovar glaciers was only 23 degrees.

4.3 Distribution and changes in glacier size

The present study mapped glaciers larger than or equal to 0.02 km^2. Milam Glacier (G080074E30524N), covering an area of 49.5 km^2 in 2020, is the largest glacier in the KSL. It is one of the major sources of the Goriganga River, which is a major tributary of the Kali River. For the analysis, glaciers were divided into five size classes: Class 1 (<0.5 km^2), Class 2 (0.05–1 km^2), Class 3 (1–5 km^2), Class 4 (5–10 km^2), and Class 5 (> = 10 km^2). The distribution of glaciers in each glacier size class showed an inverse relationship between number and area coverage. The number of glaciers in class 1 was the highest, whereas the area coverage was relatively less compared to other size classes. Almost 77% of glaciers were of size less than 0.5 km^2 (Class 1), but they covered only 18.5% of the total glacier area in the KSL. In contrast, class 3 consisted of only 11% of glaciers, covering more than 1/3 (36%) of the glacier area in the KSL. A similar pattern was also observed in each sub-basin in the KSL.

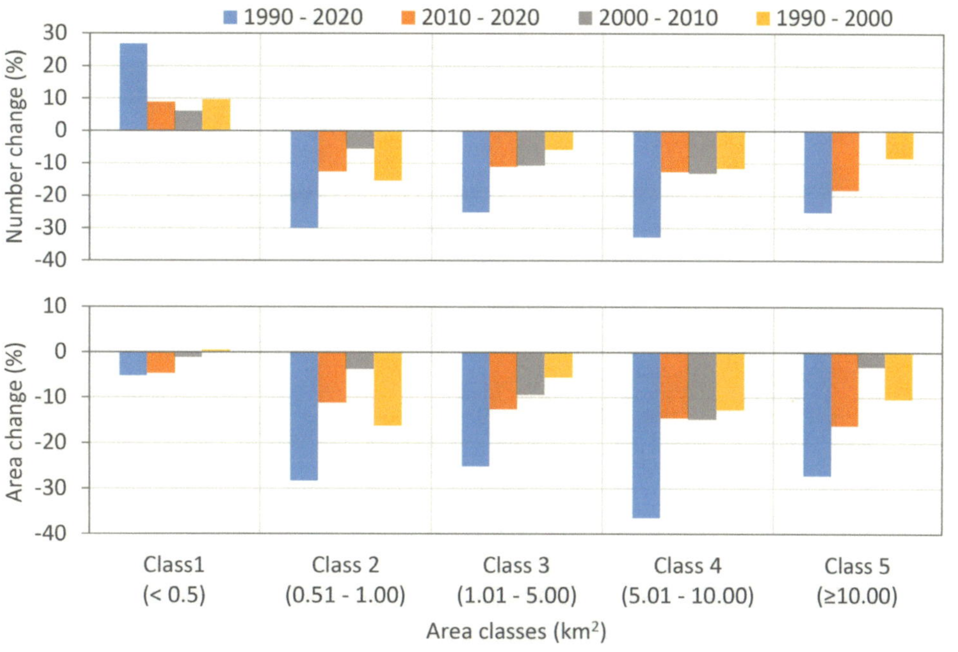

Figure 6.
Changes in number and area of glaciers in various glacier size.

The size of glaciers strongly affected changes in the number and area of glaciers in the region. The shrinkage of larger glaciers fragmented to smaller glaciers was indicated by an increment in the glacier number of class 1 size and a decrease in other size classes (**Figure 6**). The number of smaller glaciers (Class 1) increased in each decade, whereas larger glacier numbers decreased. However, the glacier area of all size classes decreased in each decade. Comparatively, changes in areas of larger glaciers were higher than in smaller glaciers. The maximum glacier area loss (33%) was in class 4 from 1990 to 2020, and glaciers larger than 10 km^2 (class 5) showed a 25% loss in 30 years. The size of the largest glacier in the region also decreased by more than 9% (54.5 km^2 in 1990 to 49.5 km^2 in 2020).

The mean glacier area also decreased from 0.87 to 0.6 km^2 from 1990 to 2020, indicating that the glaciers in the KSL are retreating, shrinking and fragmented, and the retreat rate has significantly increased in recent decades.

4.4 Terrain characteristics of glaciers and their shrinkage

The glaciers in the landscape were distributed from 3500 to 7650 masl in which the highest elevation glaciers were found in the Humla-Karnali basin and the lowest elevation glaciers from the Kali sub-basin. The glaciers in the Manasarovar basin are distributed in a narrow elevation zone from 5262 to 6717 masl. The hypsographic distribution of glaciers in the region showed that 88% of the glacier area was concentrated within 4500 to 6000 masl elevation, with the highest concentration (more than 40%) between 5000 to 5500 masl (**Figure 7**).

Glacier area in all the elevation zones decreased with the maximum glacier area changes at an elevation range from 5000 to 5500 masl (**Figure 7**). Glacier areas have not changed significantly above 6500 masl and below 4000 masl, although the area

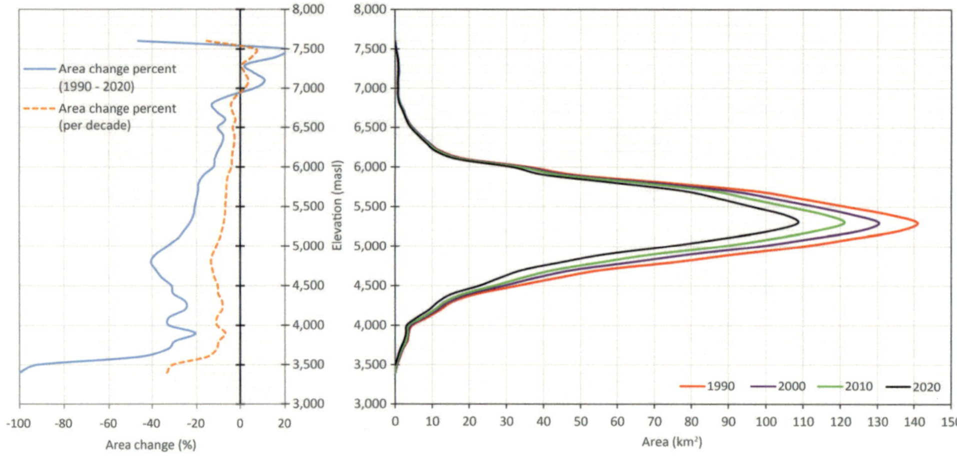

Figure 7.
Distribution and retreat of glaciers at various elevation zones from 1990 to 2020 in KSL.

percentage change rate showed that the maximum area decrease was below 4000 masl and increased above 7000 masl. The decreased area below 4000 masl indicated the retreat of the glacier terminus and upward shifting of glacier elevation, whereas the increased area in higher elevation was due to permanent snow deposition. However, the glacier area in these elevation ranges was less than 1 km^2. Glaciers below 3500 masl disappeared completely from 2010 onward. The lowest elevation of the glacier moved up from 3438 masl in 1990 to 3509 masl in 2000, 3532 masl in 2010, and 3577 in 2020. This indicated that glaciers have retreated by more than 100 m in 30 years.

The lowest-elevation glacier is highly sensitive to temperature, thereby influencing higher rates of ablations. The glacier retreat percent in lower elevations is higher, as shown in **Figure 7**. The loss percent below 3700 masl is greater than 30% in 30 years, and greater than 25% is at the elevation zone between 4500 and 5100 masl from 1990 to 2020. The area loss percentage at the highest concentration zone from 5000 to 5500 masl ranges from 20 to 40% at each 100 m elevation zone. The loss percent above 5500 masl and below 7000 masl is less than 20% in each 100 m elevation zone. Above 7000 masl, the area has increased by 9 to 20% in each 100 m elevation with an area converging of less than 1 km^2.

The slope of the glaciers also plays an important role in glacier retreat. Glaciers cannot exist in slopes greater than 60 degrees. The majority of glaciers in the KSL have mean slopes of 12 to 31 degrees and, most commonly, less than 26 degrees. Most CI glaciers lie in the upper reaches of the glaciers and attain higher mean slopes than DC glaciers. There is a clear demarcation between the mean slopes of CI and DC glaciers. CI glaciers in the KSL have an average slope above 20 degrees (ranging between 22 and 31 degrees), and DC glaciers have an average slope below 15 degrees (ranging between 12 and 13 degrees).

No significant changes were observed in the average slope of the glaciers or in CI and DC glaciers from 1990 to 2020. The distribution and retreat of the glacier area show the highest changes between slopes of 20 and 30 degrees. More than 50% of the total area changes within the 30-year period, as well as in the decadal period, was observed in this average slope range. Similarly, the second highest is from the 10 to 20 average slopes of the glaciers. This indicates that the major retreat rate is in CI glaciers in the region, and debris insulated the melting of clean ice. The number and area of glaciers of average slope range below 10 degrees has increased, indicating glacier fragmentation and shrinkage.

The glacier aspect is another critical factor that affects the melting of glacier ice. The north-facing glaciers experience less incident solar radiation and a less energetic melt regime [46]. The glaciers in KSL are distributed mostly in the southern faces with the highest number of glaciers in the southeast and southwest aspect. The area coverage of glaciers is highest in the southeast and south, indicating that most large glaciers face the southern aspect. Less than 10% of glaciers are in the northern faces, in which 3 percent of glaciers are facing the northern aspect.

The rate of glacier shrinkage is also higher in the southern aspect. More than 27 percent of the total shrinkage area was in the south in the 30-year period. Similarly, more than 21 percent of glaciers have retreated in the southeast and southwest aspects. The glacier shrinkage is significantly less in the northern aspect, with less than 1 percent of total area shrinkage in this region. The shrinkage pattern is similar for all times as well as in all basins.

5. Discussion

The present study provides comprehensive information about glacier status and its shrinkage from 1990 to 2020 in the KSL. The study adopted a semi-automatic object-based image classification method to generate the base datasets of glaciers in the KSL. Furthermore, manual intervention on shadow and low contrast difference between the glacier boundary and adjacent terrain along the lateral and distal part of glacier boundary increases the accuracy of glacier boundary. Also, to reduce mapping time and neglect discrepancies in the glacier boundary in each time period, the time series glacier data was prepared by manual editing of base data overlaying on the respective images of the years. The manual editing has some impact on individual glacier area, which is rectified by 2% while smoothing the glacier outlines [13], and the uncertainty of ±1 pixels was considered for the position of its outline. Overall, the dataset was prepared using the same spatial and spectral resolution with a consistent method and approach supported by visual interpretation, which provides reliable information on glacier changes over a 30-year period.

The study supports understanding spatial patterns of glacier characteristics such as glacier area, size, types, and topographic area variation and its changes over time throughout the landscape. The study clearly reveals the rapid shrinkage of glaciers in the landscape, with a loss of 25% of its glacier area in three decades. The decadal glacier change data shows the rate of glacier area loss increased to 1.2% per year between 2010 and 2020 from 0.83% per year between 1990 and 2000, indicating the acceleration of glacier area loss in recent decades. This rate of area loss varies in each basin and on individual glaciers, depending upon its size, type, and topographic positions. The difference between glacier area loss and its size indicates that area loss by smaller glaciers ($<0.5\ km^2$) is nominal in comparison to the area loss by bigger glaciers, but comparably, the actual glacier area loss proportion of each smaller glacier is higher. Some small glaciers have disappeared completely, and some have reduced their size to less than the mapping threshold of $0.02\ km^2$, which is not considered in this study. Nearly 5.2% of the glacier area was lost from the glacier size of $<0.5\ km^2$ between 1990 and 2020 due to the melting and shrinkage of smaller glaciers influenced by climate change. However, the glacier number has increased by 27% due to the retreat or fragmentation of glaciers larger than $0.5\ km^2$. In contrast, both the number and area of glaciers larger than 0.5 km^2 have decreased. This pattern has also been observed in other parts of the region [5, 12, 13, 24, 25], indicating that the number of smaller glaciers will increase in the future with the shrinking and fragmentation of larger glaciers.

The topographic characteristics of the glaciers also show their significant influence on glacier changes. The glacier area loss in the lower elevation zones is distinctive as the lowest elevation of the glacier has moved from 3429 masl in 1990 to 3577 masl in 2020. Glaciers below 3500 masl vanished completely after 2010, and more than 22% of the area was lost at elevations below 4000 masl. However, the surface area loss is very low, below 4000 masl, compared to losses at elevations from 4500 to 5500, which is the highest area loss in all decades. In contrast, the glacier area has slightly increased above 7000 masl, indicating snow accumulation and the formation of permanent ice on mountain slopes. Other topographic characteristics such as slope and aspect also play prominent roles in the existence of glaciers and their area changes. Low-slope glaciers are more susceptible to area changes in response to climate change. In KSL, glacier slopes mainly range up to 50 degrees, with a few portions of glacier area above 50 degrees. Proportionally, area change is higher in high-degree sloped glaciers, whereas the area loss and coverage are higher between 20 and 30 degrees. Similarly, the distribution and area loss is higher in the southern aspect of glaciers.

A reduction in CI glacier area has been observed from 1990 to 2020 as these types of glaciers respond quickly to a warming environment in a shorter period than debris-covered glaciers. The increased area percentage of debris-covered glaciers over 30 years indicated that the landscape has less precipitation and high temperature, which rapidly increased the glacier melting process, thinning the glacier fronts, CI to be exposed to bare rock and/or becoming a part of debris glaciers. Sometimes, a rapid recession can also detach CI from DC because of the steepness of the topography or vertical cliff over which the glacier has been passing. In addition, the pace of recession on DC glaciers can form supraglacial ponds on the surface and expansion of glacial lakes attached to or near the glaciers. This suggests that detailed monitoring is required to better characterize the response of the glaciers to the surroundings in the KSL.

The landscape is also rich in religious and cultural heritage. The Mt. Kailash and adjacent Lake Manasarovar are spiritual focal places for millions of followers of Hinduism, Buddhism, Jainism, Sikhism, and Bon religions [47]. Mount Kailash is covered by a glacier ice cap (**Figure 8**), which is split into four glaciers based on their hydrological flow. The Polung and Gangjam glaciers in the north and northeast [48] have retreated 25.4 and 35.2% within a 30-year period, respectively. Overall, the glaciers on Mt. Kailash have retreated by 26.7% during the 30-year period. The shrinkage of glaciers in the mountain will expose the underlying rock and reduce the beauty and significance of this sacred mountain peak. Moreover, the Manasarovar Lake is also fed by several glaciers around the valley. Overall, the glacier area around the Manasarovar inland valley decreased by 27% during the study period. The recession of glacier mass, in the long run, will reduce meltwater supply into the lake, which will reduce the size of lakes and other water bodies and impact their water quality [49, 50]. In many religions, particularly Hinduism and Buddhism, water bodies are believed to hold purifying and cleansing powers and are essential for ceremonies and prayers [51]; therefore, changes in water quality and quantity due to glacial changes are likely to have implications on cultural practices associated with water bodies in the landscape.

Shrinkage in glaciers increases uncertainties in water supplies as well as disaster risks, further exacerbating the vulnerability of the mountain communities to climate change, threatening water, food, and energy security in upstream and downstream areas of the landscape. Initially, the rapid melting of glaciers might lead to an increase in

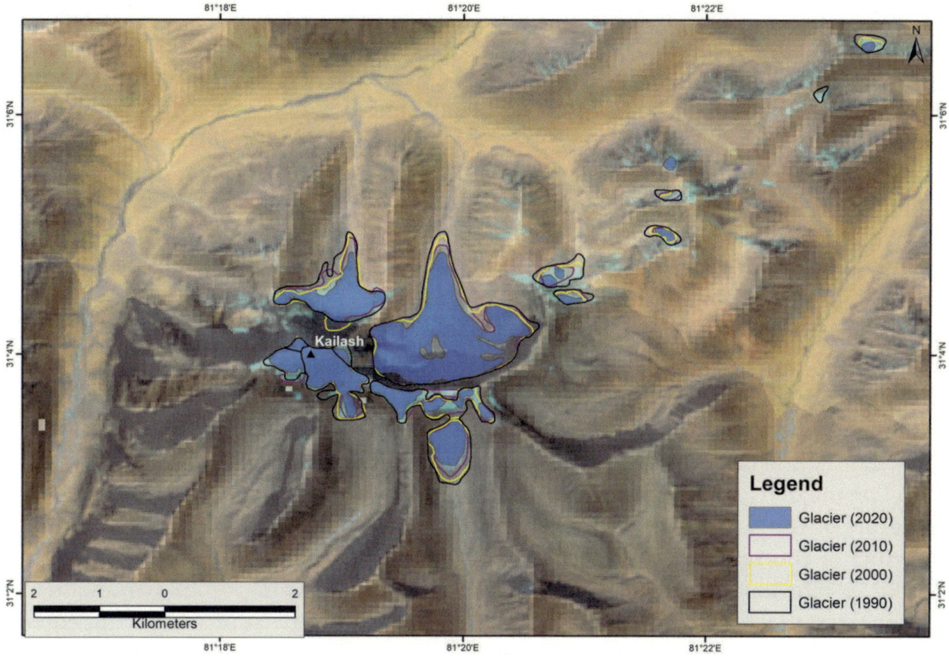

Figure 8.
Shrinking and retreating glaciers in the Mt. Kailash region from 1990 to 2020.

glacier melt runoff and the river flow lasting some decades, but with the gradual reduction in glacier mass, the river flow will also decline in the long term [52]. This alteration in the contribution of glacier meltwater to the rivers will have adverse impacts on the ecosystem services, affecting upstream as well as downstream populations [53, 54]. Many hydropower and irrigation systems such as Dhauliganga Hydropower (230 MW) in India, Chameliya Hydropower (30 MW) in Nepal, and many traditional and modern irrigation systems in the upstream and downstream areas of the landscape play an important role in agriculture and food production, will be impacted in the long run.

Rapid glacier recession also increases the formation and expansion of glacial lakes especially below the retreating terminus. This increases the risks of glacial lake outburst flood (GLOF) events, threatening lives, livelihood, and property, including infrastructures and cultural heritage. The recurring GLOF event in the Limi Valley of Humla District in Nepal has threatened the settlement, as well as one of the oldest Buddhist monasteries in the country [32]. Thus, there is an urgent need to take appropriate adaptive and mitigation measures to reduce further acceleration of risks. Though overall trends of decline in glaciers are clear, factors influencing the susceptibility of glaciers to changes, such as type, size, aspect, slope, elevation, proximity to water bodies, and presence of thick or thin layers of debris, vary. Detailed information on these characteristics and variations in response is needed to support accurate projections of future glacier change under different climate scenarios using glacier modeling.

6. Conclusion

The present datasets of glaciers from 1990 to 2020 revealed that the glaciers have retreated significantly. Due to the shrinking, retreating, and fragmenting of individual glaciers, the number of glaciers had increased, and the area decreased. Overall, the glacier area has decreased by 25.5%, and the number of glaciers has increased by 7.8% (141) over the period of 30 years. The loss in glacier area was only 15% in 20 years from 1990 to 2010, but an additional loss of 11% occurred in 10 years between 2010 and 2020, indicating the recent loss of glaciers was almost double that of the previous decades. Decreases in glacier area accompanied by an increase in the number of glaciers are clear evidence of fragmentation because of uneven shrinkage of individual glaciers. Therefore, the number of smaller glaciers increased, while the glacier area of glaciers of all sizes decreased. Small glaciers and low-elevation glaciers are more sensitive to climate change and melt faster. Some of the smaller glaciers present in 1990 had disappeared by 2010, and so on. However, the retreat rate is higher in lower elevations, and the glaciers below 3500 masl completely disappeared in 2010. The distribution of glacier area and its retreat amount was higher at elevations ranging from 5000 to 5500 masl. Moreover, glaciers on steep slopes facing the southern aspect and frontal parts of the glaciers associated with glacial lakes are retreating faster. The majority of the glaciers in KSL are facing the southern aspect, and the retreat rate of the glaciers in the south, southwest, and southwest aspect is significantly higher (more than 20 percent in SE and around 50% in S and SW).

Separate information for CI and DC glaciers will serve as an important parameter in climate change models. The reduction of CI glacier area by 29 percent and the increment of DC glaciers by 13 percent within 30 years indicates a distinct response to climate warming.

The present study concluded that the glacier system is very complex, and spatial changes vary depending on its topographic setup in terms of altitude, slope, aspect, and composition. However, the study shows that the glaciers in the region are steadily depleting, and its trend of retreat will continue in the future with a changing climate. However, the present increase in the melting of glaciers might enhance economic opportunities and productivity of the region by surplus supply of freshwater for livelihood, agriculture, and hydropower generation. It is also necessary to understand the availability of these water resources, which will be depleted with the reduction of glaciers in the long run. At the same time, some of the glaciers in the region represent cultural sites. Each year, thousands of pilgrims visit these places, which signifies the importance of glaciers. Hence, the systematic monitoring and observed changes must be assessed in the long run, and these data can be utilized for future projections of water resource availability using various climatic and glacio-hydrological models. Also, these datasets are very useful for planning and management of water resources and for disaster risk reduction.

Acknowledgements

We are indebted to Janita Gurung and Miriam Jackson for their full support in this work. Thank you to the following Tribhuvan University staff – Kabiraj Paudyal & Ramita Bajracharya from the Central Department of Geology, Mamata Sayami & Subash Acharya from the Department of Geology, Trichandra Multiple College, and Anita Tuitui & Earina Sthapit from the Central Department of Hydrology and Meteorology for support during preparation and mapping of glaciers. The study was supported by the Cryosphere intervention (previously Cryosphere initiative) and Kailash Scared Landscape Conservation and Development Initiative in ICIMOD. Landsat data are courtesy of NASA and the USGS. The SRTM elevation model version is courtesy of NASA's Jet Propulsion Laboratory and was further processed by the Consultative Group for International Agriculture Research. ICIMOD gratefully acknowledges the support of its core donors: the Governments of Afghanistan, Australia, Austria, Bangladesh, Bhutan, China, India, Myanmar, Nepal, Norway, Pakistan, Sweden, and Switzerland.

Conflict of interest

The authors declare no conflict of interest.

Disclaimer

The views and interpretations in this publication are those of the authors. They are not necessarily attributable to ICIMOD and do not imply the expression of any opinion by ICIMOD concerning the legal status of any country, territory, city, or area of its authority, or concerning the delimitation of its frontiers or boundaries, or the endorsement of any product.

Author details

Sudan Bikash Maharjan*, Tenzing Choyal Sherpa, Finu Shrestha and Binaya Pasakhala
ICIMOD, Kathmandu Nepal

*Address all correspondence to: sudan.maharjan@icimod.org

References

[1] Drew G. A retreating goddess? Conflicting perceptions of ecological change near the Gangotri-Gaumukh glacier. Journal for the Study of Religion, Nature, and Culture. 2012;**6**:3. DOI: 10.1558/jsrnc.v6i3.344

[2] Knight PG. Glacier: Nature and Culture. London: Reaktion Books; 2019

[3] Huss M, Bookhagen B, Huggel C, Jacobsen D, Bradley RS, Clague JJ, et al. Toward mountains without permanent snow and ice. Earth's Future. 2017;**5**:418-435. DOI: 10.1002/2016EF000514

[4] Tuladhar S, Pasakhala B, Maharjan A, Mishra A. Unravelling the linkages of cryosphere and mountain livelihood systems: A case study of Langtang, Nepal. Advances in Climate Change Research. 2021;**12**(1):119-131. DOI: 10.1016/j.accre.2020.12.004

[5] Bajracharya SR, Maharjan SB, Shrestha F, Guo W, Liu S, Immerzeel W, et al. The glaciers of the Hindu Kush Himalayas: Current status and observed changes from the 1980s to 2010. International Journal of Water Resources Development. 2015;**31**(2):161-173. DOI: 10.1080/07900627.2015.1005731

[6] Armstrong R, Raup B, Khalsa SJS, Barry R, Kargel J, Kieffer H. GLIMS Glacier Database. Boulder CO, USA: National Snow and Ice Data Center; 2011

[7] Pfeffer WT, Arendt AA, Bliss A, Bolch T, Cogley JG, Gardner AS, et al. The Randolph consortium. The Randolph glacier inventory: A globally complete inventory of glaciers. Journal of Glaciology. 2014;**60**(221):537-552. DOI: 10.3189/2014JoG13J176

[8] Nuimura T, Sakai A, Taniguchi K, Nagai H, Lamsal D, Tsutaki S, et al. The GAMDAM glacier inventory: A quality-controlled inventory of Asian glaciers. The Cryosphere. 2015;**9**(3):849-864. DOI: 10.5194/tc-9-849-2015

[9] Guo W, Liu S, Xu J, Wu L, Shangguan D, Yao X, et al. The second Chinese glacier inventory: Data, methods, and results. Journal of Glaciology. 2015;**61**(226):357-372. DOI: 10.3189/2015JoG14J209

[10] Jiang S, Nie Y, Liu Q, Wang J, Liu L, Hassan J, et al. Glacier change, supraglacial debris expansion and glacial lake evolution in the Gyirong River basin, Central Himalayas, between 1988 and 2015. Remote Sensing. 2018;**10**:986

[11] Bajracharya SR, Mool PK, Shrestha BR. Impact of Climate Change on Himalayan Glaciers and Glacial Lakes- Case Studies on GLOF and Associated Hazards in Nepal and Bhutan. Kathmandu: ICIMOD; 2007. DOI: 10.53055/ICIMOD.470

[12] Bajracharya SR, Maharjan SB, Shrestha F, Bajracharya OR, Baidya S. Glacier status in Nepal and decadal change from 1980 to 2010 based on landsat data. ICIMOD; 2014;**2**:1-60. DOI: 10.53055/ICIMOD.591

[13] Bajracharya SR, Maharjan SB, Shrestha F. The status and decadal change of glaciers in Bhutan from 1980's to 2010 based on the satellite data. Annals of Glaciology. 2014;**55**(66):159-166. DOI: 10.3189/2014AoG66A125

[14] Ives JD, Shrestha RB, Mool PK. Formation of glacial lakes in the Hindu-Kush Himalayas and GLOF risk assessment. ICIMOD. 2010:1-44. DOI: 10.53055/ICIMOD.521

[15] Bolch T, Kulkarni A, Kääb A, Huggel C, Paul F, Cogley JG, et al. The state and fate of Himalayan glaciers. Science. 2012;**336**(6079):310-314. DOI: 10.1126/science.1215828

[16] Mernild SH, Lipscomb WH, Bahr DB, Radic V, Zemp M. Global glacier changes: A revised assessment of committed mass losses and sampling uncertainties. The Cryosphere. 2013;**7**:1565-1577. DOI: 10.5194/tc-7-1565-2013

[17] Gardent M, Rabatel A, Dedieu JP, Deline P. Multitemporal glacier inventory of the French Alps from the late 1960s to the late 2000s. Global and Planetary Change. 2014;**120**:24-37. DOI: 10.1016/j.gloplacha.2014.05.004

[18] Wester P, Mishra A, Mukherji A, Shrestha AB. The Hindu Kush Himalaya Assessment—Mountains, Climate Change, Sustainability and People. Cham: Springer Nature Switzerland AG; 2019

[19] IPCC. The Physical Basis: Contribution of Working Group I to the Fifth Assessment Report of the Intergovernmental Panel on Climate Change. Cambridge, UK; New York, NY, USA: IPCC; 2013

[20] Kraaijenbrink P, Bierkens M, Lutz A, Immerzeel WW. Impact of a global temperature rise of 1.5 degrees Celsius on Asia's glaciers. Nature. 2017;**549**:257-260. DOI: 10.1038/nature23878

[21] Bolch T, Pieczonka T, Benn DI. Multi-decadal mass loss of glaciers in the Everest area (Nepal Himalaya) derived from stereo imagery. The Cryosphere. 2011;**5**:349-358. DOI: 10.5194/tc-5-349-2011

[22] Yao TD, Thompson L, Yang W, Yu WS, Gao Y, Guo XJ, et al. Different glacier status with atmospheric circulations in Tibetan plateau and surrounding. Nature Climate Change. 2012;**2**:663-667. DOI: 10.1038/NCLIMATE1580

[23] Xiang Y, Gao Y, Yao T. Glacier change in the Poiqu River basin inferred from Landsat data from 1975 to 2010. Quaternary International. 2014;**349**:393-401. DOI: 10.1016/j.quaint.2014.03

[24] Maharjan SB, Shrestha F, Azizi F, Joya E, Bajracharya B, Bromand MT, et al. Monitoring of glaciers and Glacial Lakes in Afghanistan. In: Bajracharya B, Thapa RB, Matin MA, editors. Earth Observation Science and Applications for Risk Reduction and Enhanced Resilience in Hindu Kush Himalaya Region. Cham: Springer; 2021. DOI: 10.1007/978-3-030-73569-2_11

[25] Maharjan SB, Joya E, Rahimi MM, Azizi F, Muzafari KA, Bariz M, et al. Glaciers in Afghanistan: Status and changes from 1990 to 2015. Research Report. ICIMOD and NWARA. 2021. DOI:10.53055/ICIMOD.784.

[26] Maurer JM, Schaefer JM, Rupper S, Corley A. Acceleration of ice loss across the Himalayas over the past 40 years. Science Advances. 2019;**5**:6. DOI: 10.1126/sciadv.aav7266

[27] Bhambri R, Bolch T, Kawishwar P, Dobhal DP, Srivastava D, Pratap B. Heterogeneity in glacier response in the upper Shyok valley, Northeast Karakoram. The Cryosphere. 2013;**7**(5):1385-1398. DOI: 10.5194/tc-7-1385-2013

[28] Minora U, Bocchiola D, D'Agata C, Maragno D, Mayer C, Lambrecht A, et al. Glacier area stability in the Central Karakoram National Park (Pakistan) in 2001-2010: The "Karakoram anomaly" in the spotlight. Progress

in Physical Geography: Earth and Environment. 2016;**40**(5):629-660. DOI: 10.1177/0309133316643926

[29] Bolch T, Buchroithner M, Pieczonka T, Kunert A. Planimetric and volumetric glacier changes in the Khumbu Himal, Nepal, since 1962 using Corona, Landsat TM and ASTER data. Journal of Glaciology. 2008;**54**(187):592-600. DOI: 10.3189/002214308786570782

[30] Azam MF, Wagnon P, Berthier E, Vincent C, Fujita K, Kargel JS. Review of the status and mass changes of Himalayan-Karakoram glaciers. Journal of Glaciology. 2018;**64**:61-74. DOI: 10.1017/jog.2017.86

[31] Lamsal D, Sawagaki T, Watanabe T. Digital terrain modelling using Corona and ALOS PRISM data to investigate the distal part of Imja glacier, Khumbu Himal, Nepal. Journal of Mountain Science. 2011;**8**:390-402. DOI: 10.1007/s11629-011-2064-0

[32] Kropáček J, Neckel N, Tyrna B, Holzer N, Hovden A, Gourmelen N, et al. Repeated glacial lake outburst flood threatening the oldest Buddhist monastery in North-Western Nepal. Natural Hazards and Earth System Sciences. 2015;**15**(10):2425-2437. DOI: 10.5194/nhess-15-2425-2015

[33] Jha LK, Khare K. Glacial Lake outburst flood (GLOF) study of Dhauliganga basin in the Himalaya. Peter Christian Speldewinde (reviewing editor) cogent. Environmental Science. 2016;**2**:1-13. DOI: 10.1080/23311843.2016.1249107

[34] Zomer RJ, Trabucco A, Metzger M, Oli KP. Environmental stratification of Kailash sacred landscape and projected climate change impacts on ecosystems and productivity. In: ICIMOD Working Paper, ICIMOD. 2013. p. 1. DOI: 10.53055/ICIMOD.578

[35] ICIMOD. Transboundary Landscapes: Kailash Sacred Landscape Conservation and Development Initiative. 2020. Available from: https://www.icimod.org/initiative/ksl

[36] Bajracharya SR, Shrestha B. The Status of Glaciers in the Hindu Kush-Himalayan Region. Kathmandu: ICIMOD; 2011. DOI: 10.53055/ICIMOD.551

[37] Maharjan SB, Shrestha F, Bajracharya SR. Training manual on application of remote sensing and geographic information systems for mapping and monitoring of glaciers, introduction to geographic systems (GIS) and ArcGIS. ICIMOD Manual. 2017/11. DOI: 10.53055/ICIMOD.702

[38] Bajracharya SR, Maharjan SB, Shrestha F. Training manual on application of remote sensing and geographic information systems for mapping and monitoring of glaciers using eCognition software. ICIMOD Manual. 2017/10. DOI: 10.53055/ICIMOD.701

[39] Stokes CR, Shahgedanova M, Evans IS, Popovnin VV. Accelerated loss of alpine glaciers in the Kodar Mountains, South-Eastern Siberia. Global and Planetary Change. 2013;**101**:82-96. DOI: 10.1016/j.gloplacha.2012.12.010

[40] DeBeer CM, Sharp MJ. Recent changes in glacier area and volume within the southern Canadian cordillera. Annals of Glaciology. 2007;**46**:215-221. DOI: 10.3189/172756407782871710

[41] Bolch T, Menounos B, Wheate R. Landsat-based inventory of glaciers in western Canada, 1985-2005.

Remote Sensing of Environment. 2010;**114**(1):127-137. DOI: 10.1016/j.rse.2009.08.015

[42] Frey H, Paul F, Strozzi T. Compilation of a glacier inventory for the western Himalayas from satellite data: Methods, challenges, and results. Remote Sensing of Environment. 2012;**124**:832-843. DOI: 10.1016/j.rse.2012.06.020

[43] Mattson LE, Gardner JS, Young GJ. Ablation on debris covered glaciers: An example from the Rakhiot glacier, Panjab, Himalaya. In: Snow and Glacier Hydrology (Proceeding of the Kathmandu Symposium, November 1992). Vol. 218. IAHS Publication; 1993

[44] Reznicheko N, Davies T, Shulmeister J, McSaveney M. Effects of debris on ice-surface melting rates: An experiment study. Journal of Glaciology. 2010;**56**(197). DOI: 10.3189/002214310792447725

[45] Immerzeel WW, Kraaijenbrink PDA, Shea JM, Shrestha AB, Pellicciotti F, Bierkens MFP, et al. High resolution monitoring of Himalayan glacier dynamics using unmanned aerial vehicles. Remote Sensing Environment. 2014;**150**:93-103. DOI: 10.1016/j.rse.2014.04.025

[46] Marshall S, White E. Alberta Glacier Inventory and Ice Volume Estimation. Canmore: Crowfoot Ice Research and Consulting; 2010. Available from: https://albertawater.com/dmdocuments/01_2010_12_Alberta_Glacier_Inventory_and_Ice_Volume_Estimation_Marshall_et_al%20(15).pdf

[47] Pandey A, Kotru R, Pradhan N. Kailash sacred landscape: Bridging cultural heritage, conservation, and development through a transboundary landscape approach. In: Verschuuren B, Furuta N, editors. Asian Sacred Natural Sites. Routledge; 2016. pp. 145-158

[48] Juyal N, Thakkar PS, Sundriyal YP. Geomorphic evidence of glaciation around mount Kailash (inner Kora): Implication to past climate. Current Science. 2011;**100**(4) http://www.jstor.org/stable/24073091

[49] Sharma BM, Nizzetto L, Bharat GK, Tayal S, Melymuk L, Sanka O, et al. Melting Himalayan glaciers contaminated by legacy atmospheric depositions are important sources of PCBs and high-molecular-weight PAHs for the Ganges floodplain during dry periods. Environmental Pollution. 2015;**206**:588-596. DOI: 10.1016/j.envpol.2015.08.012

[50] Sun X, Wang K, Kang S, Guo J, Zhang G, Huang J, et al. The role of melting alpine glaciers in mercury export and transport: An intensive sampling campaign in the Qugaqie basin, inland Tibetan plateau. Environmental Pollution. 2017;**220**:936-945. DOI: 10.1016/j.envpol.2016.10.079

[51] Singh RPB. Water symbolism and sacred landscape in Hinduism: A study of Benares (Varasani). Erdkunde. 1994;**48**(3):210-227. DOI: 10.3112/erdkunde.1994.03.05

[52] IPCC. Climate change 2014- impact, adaptation and vulnerability. In: Part B: Regional Aspects. Contribution of Working Group II to the Fifth Assessment Report of the Intergovernmental Panel on Climate Change. IPCC: Cambridge Univ. Press, USA; 2014

[53] Milner AJ, Kahmis K, Battin TJ, Brittain JE, Barrand NE, Fureder L, et al. Glacier shrinkage driving global changes in downstream systems. Proceedings of the National Academy of Sciences of the United States of America. 2017;**114**:9770-9778

[54] Mukherji A, Sinisalo A, Nüsser M, Garrard R, Eriksson M. Contributions

of the cryosphere to high mountain communities of the Hindu Kush Himalaya: A review. Regional Environmental Management. 2019:127-154, 211-245. DOI: 10.1007/s10113-019-01484-w

Section 4

Glaciers, Precipitation and Water Use

Chapter 5

Climate Change in Tropical Glacier-Fed Rivers, Contrasting Global Responses and Future Implications on Stream Functional Diversity

Patricio Andino Guarderas and Rodrigo Espinosa Barrera

Abstract

Glaciers in the tropics have unique characteristics, such as their melting patterns and the impact of glacial influence on our environment. This chapter's objective is to assess melting tropical glaciers effects on the ecological dynamics of highland glacier-influenced streams, including changing local patterns due to climate change. Life traits such as trophic interactions and phenology patterns of a multi environmental riverine system of distinct origins are inquired. These diverse patterns alter some ecosystem biotic functions such as resilience and decomposition of highland rivers. Population and community approaches, perspectives on ecosystem function, and future impacts are included, as tropical glacier's melting effect changes and diminishes in time. The density, richness, and composition of macroinvertebrate assemblages in these areas are influenced by factors such as altitude, glacier coverage, conductivity, temperature, and channel stability. Dispersal plays a crucial role in shaping the succession of glacier forelands and similar environment, as spatial beta-diversity is influenced by nestedness and turnover, indicating distinct mechanisms driving diversity. Climate change is bound to affect glacier catchments and downstream aquatic ecosystems, not only by decreasing quantity but quality of water; projected higher human demand and lower water supply might cause future local conflicts over water availability for human use.

Keywords: tropical glaciers, high altitude, aquatic ecosystems, environmental harshness, functional diversity, climate change, glacier melting

1. Introduction

Tropical glaciers are not globally common as they only form when the ground altitude exceeds 5000 m a.sl. Although they do occur in some continents of the Earth, tropical regions are distinguished because of their lack of temperate seasons and its increased exposure to the sun, which gives special vulnerability to glaciers over climatic variations [1]. Thus, the biological and ecological dynamics of tropical highland glacier-fed or influenced ecosystems might be affected in the future, distressing ecological processes linked with water quality and human use. To point out the basic features of tropical glacier, we will begin with a case well studied by the authors of this chapter, the ecology of tropical glacier-fed streams in the Ecuadorian Andes which differs from the traditional model based on temperate glacier-fed streams. Key findings in Ecuador may apply to other tropical glacier feed streams including differences between the composition and distribution of macroinvertebrate communities, the lack of a classical kryal zone (where temperatures are below 4°C) and the influence of stream temperature and physical stability on faunal composition and richness. As there is little seasonality in air temperature, the night time temperature is always below freezing, just as shown in **Figure 1**, also indicating that daily diel temperature variation is greater in the tropics along the whole year than its diel variation on each season in a temperate region glacier. Therefore, melting of the glacier virtually stops during night and early morning. For this reason, the intense insolation in the morning from the quickly rising equatorial sun warms up this small volume of water to a high temperature, except when it is foggy and cold. As **Figure 2** describes, later during the day, melting increases, and the larger volume of ice-cold melt water brings temperature down again in early afternoon [2].

Tropical glaciers located in regions like the Himalayas, Andes, and Altiplano exhibit unique dynamics compared to temperate glaciers, and so they have important hydrological and ecological implications [3]. Aerial vehicle surveys in Peru have

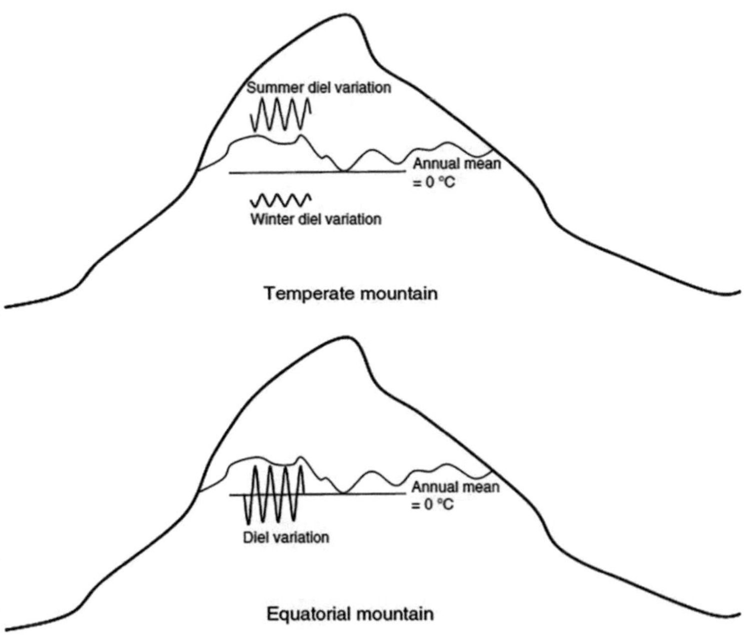

Figure 1.
Conceptual diagram of a tropical glacier illustrating air temperature variations at a temperate and equatorial glacial terminus (both located at the altitude of annual Tmean = 0°C). For the temperate mountain, the altitude at which the summer Tmean is 0°C is higher than the glacial terminus, while that where the winter Tmean is 0°C, the altitude is lower. For equatorial mountain, Tmax during the day at the glacial terminus is >0°C throughout the year, whereas Tmin during the night is always below 0°C [2].

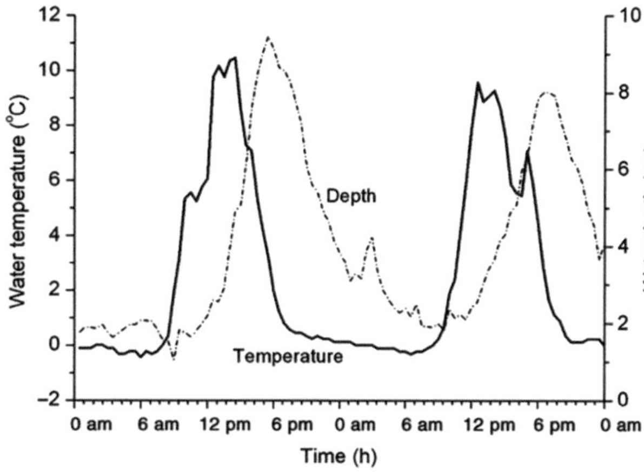

Figure 2.
Daily water thermic and flow fluctuations (depth). Water temperature (continuous line) and depth (dashed line) from site 1B (at the glacier snout) during an example 48-h period at the end of February 2008 [2].

revealed heterogeneous patterns of change within glaciers, including zones of accelerated ice loss and subsidence within proglacial lakes [4]. The Himalayan-Karakoram region of South Asia shares the same problematic as extensive glaciers and snowfields contribute meltwater to major river systems supplying water for hundreds of millions of people [5]. At the Bolivian Altiplano, where glaciers are nearly absent, ice-rich permafrost still impacts debris accumulation sectors, and its potential degradation could affect regional hydrology and consequently macrofaunal dynamics [6]. Global glaciohydrology changes could significantly impact sustainable water resource management and geopolitics, as this high altitude alteration is also expected to affect downstream more populated systems.

Fast tropical glacier retreat is altering landscapes and ecosystems [1]. The Colombian glacier "Glaciar Conejeras" has shrunk dramatically since 2006 with downstream hydrological patterns influenced by complex interactions between temperature, precipitation, and glacier shrinkage [7]. Glacial influence on Andean streams also shapes biological communities as higher glaciality influence is correlated with increased abundance of small, hardy macroinvertebrate taxa adapted to harsh conditions [8]. Other case emphasized in microbiomes showed that glacier-fed streams in Africa harbor unique bacterial assemblages compared to normal surface waters [9]. Colonization patterns following deglaciation indicate pioneer communities are dominated by generalist species with high dispersal abilities, while diversity increases over time since exposure to harsh conditions decline [10]. Functional diversity and certain functional feeding groups also seem to decline under stronger glacial influence likely due to environmental filtering even though decomposition rates remain similar to other alpine streams [11, 12], as the same abrasive physical and chemical conditions seem to drive litter decomposition in highly influenced glacier-fed streams [11]; macrofaunal abundance also decreases with greater glacial coverage in catchments [12]. Rare taxa preferentially inhabit highly disturbed areas created by glacial floods threatening their persistence as glaciers shrink [13], and this effect caused by an intermediate disturbance occurs especially during glacier flood pulses modulating both macroinvertebrate abundance and their food resources, thus allowing other species to inhabit spaces where more competitive or dominant species are not able to stabilize as this harsh environment events keep drifting individual downstream [12–14].

Weather shifts due to climate change are not a strange phenomenon in the tropics, and so tropical glaciers melting and their eventual diminishing pose a great concern to local and regional human populations, as this matter is about a vital non-renewable resource, water. Major shifts are unfolding across impacted landscapes including formation of new lakes and streams as well as adjustments to species distribution, assemblages, and available habitat. These adjustments will eventually have an impact on the ecological functionality of tropical glacier feed streams and its consequent effects on downstream hydrological systems; as tropical glaciers disappear entirely, impacts on alpine and high-elevation biota are projected to grow [12, 15]. Glacier retreat, driven by climate change is causing significant impacts on hydrological cycles and downstream ecosystems worldwide. In tropical regions, even a short-term reduction in meltwater flow can induce significant changes in benthic fauna community composition within a matter of weeks [16], and both algal and herbivore biomass increase in response to flow reduction highlighting the ecological effects of ongoing glacier retreat on aquatic systems. The disappearance of glaciers in a catchment can lead to the loss of 11–38% of regional species pools including endemic species. Moreover, the steady shrinkage of glaciers reduces taxon turnover and local richness in downstream reaches with lower glacial cover [15]. Predictable mechanisms govern river invertebrate community responses to decreasing glacier cover globally [17]. Glacier retreat affects not only aquatic but also terrestrial ecosystems with similar implications. For example, in plant diversity and community dynamics, there is an initial increase in plant diversity with glacier retreat which is followed by a decrease after glacier extinction. Approximately 22% of plant species show a non-linear response to glacier retreat and are at risk of local disappearance, as soil carbon enrichment and reduction of

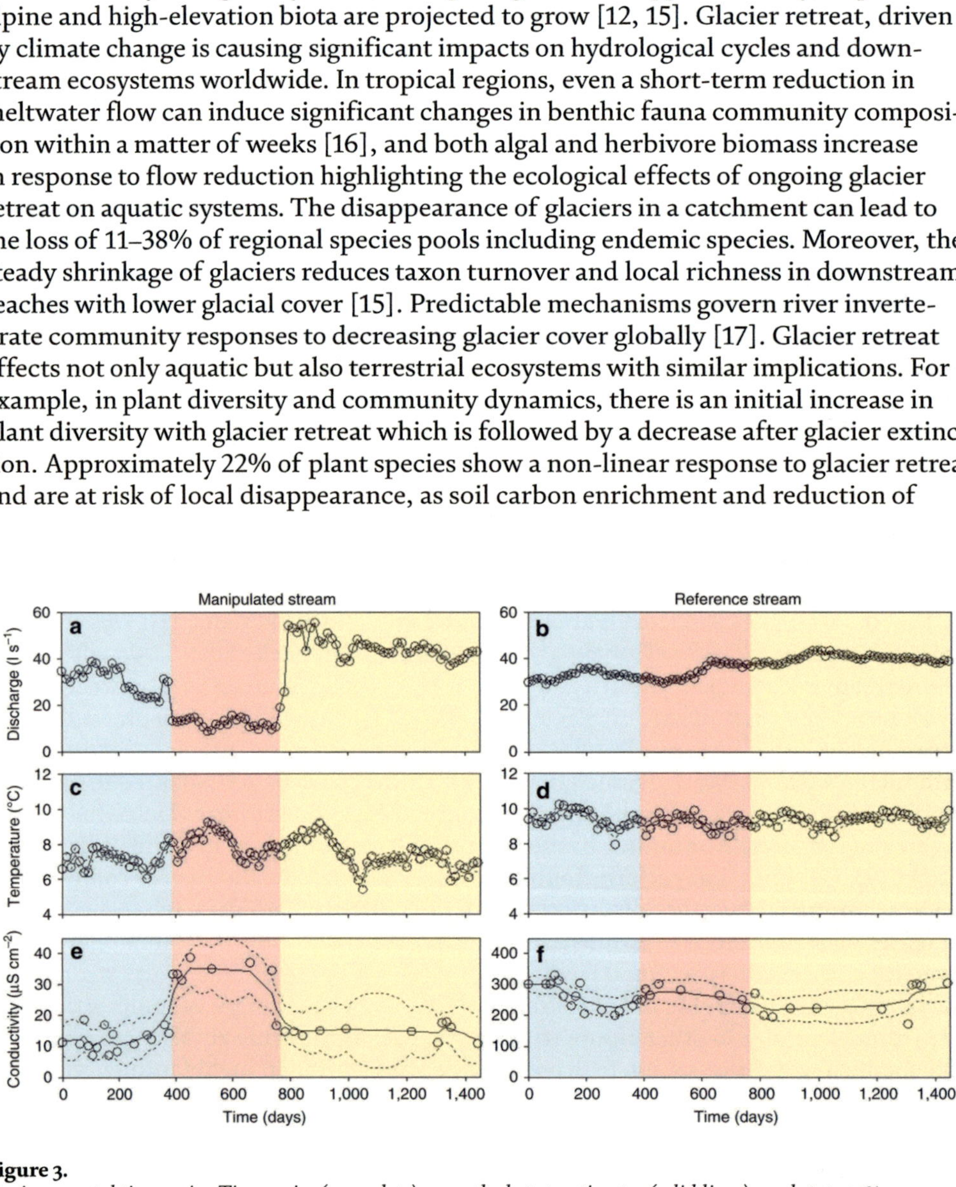

Figure 3.
Environmental time series. Time series (open dots), smoothed-state estimates (solid lines), and state 95% confidence level (dotted lines) of discharge (a, b), water temperature (c, d), and conductivity (e, f) for the downstream reach of the manipulated stream and the reference stream. Blue, red, and yellow regions indicate base flow, flow alteration, and flow recovery periods, respectively [16].

physical disturbance contribute to distribution patterns of plant species, while species associations shift from facilitation to competition [18].

While the vulnerability of river ecosystems to glacier retreat is evident in both tropical and temperate regions, some differences in ecological responses emerge, for instance, the time required for a tropical stream system to return to its initial state after flow reduction was found to be within 14 and 16 months [16]; additionally, the specific taxa and ecological processes driving the response to glacier retreat may vary between regions [19, 20]. We must take into account that even flow fluctuations may affect relevant biotic parameters. **Figure 3** shows the results from an experimental research held on a multi-stream survey of sites varying in glacial influence showing an abrupt increase in algal and herbivore biomass below 11% glacier cover in the catchment, indicating that flow reduction strongly affects glacier-fed stream biota, prefiguring profound ecological effects of ongoing glacier retreat on aquatic systems [16].

Climate change and glacier retreat have significant impacts on aquatic ecosystems in both tropical and temperate regions. Glacier-fed rivers are often used as model systems for understanding the effects of climate change on ecosystems due to their strong atmospheric-cryospheric links and high biodiversity. That is why, understanding river invertebrate's diversity patterns play a crucial role in ecosystem functions and services, and it is threatened by global change, including glacier retreat. This chapter looks to assess the ecological responses, biodiversity loss, and community dynamics associated with glacier retreat in tropical and temperate regions, providing valuable insights for conservation efforts and the protection of sensitive species.

2. Tropical glacier-fed Rivers: functional diversity under threat

2.1 Tropical glacier melting: ecological effects and differences compared to temperate regions

Glacier retreat driven by climate change is impacting aquatic ecosystems globally [1]; however, it is important to recognize that the ecological effects of glacier melting may differ between tropical and temperate regions due to variations in deglaciation rates and environmental conditions [21]; understanding these differences is crucial for assessing the impacts of climate change on tropical ecosystems and developing appropriate adaptation and mitigation measures [22, 23]. In tropical regions, the short-term reductions in glacier melt can have significant and rapid effects on benthic communities within weeks. This is largely due to the strong atmospheric-cryospheric links that exist within these regions where changes in glacier melt can directly influence various discharge parameters and in consequence the aquatic ecosystem [22]. Moreover, the complete disappearance of tropical glaciers can lead to the loss of a substantial percentage (11–38%) of regional species pools, posing severe risks to high-elevation specialist species [23]. The accelerated deglaciation in tropical regions further exacerbates these risks, as tropical glaciers are receding rapidly under climate change [22].

On the other hand, temperate glacier-fed systems demonstrate similar ecological responses to reduced glacier cover. For example, the decrease in glacier extent in both tropical and temperate regions can result in increased algal and herbivore biomass [21–23]. The decline in macroinvertebrate diversity is expected to occur with glacier shrinkage in both biomes [21, 23]; however, it is important to note that the specific taxa and processes driving these changes may vary between regions. This global analysis has revealed consistent increases in the functional diversity of benthic invertebrates with decreasing glacier cover regardless of the biome [23]. The pattern persists even after accounting for latitude, indicating that the changes are regulated by trait drive interactions like dispersal limitations and environmental filtering [21]. Underlining the complex interplay between glacier retreat, its accelerating melting and biodiversity dynamics suggest that there are common ecological responses to glacier melting across different regions. In terms of physicochemical characteristics and biodiversity, tropical glacier-fed rivers experience more pronounced changes compared to temperate regions due to the rapid deglaciation process. These variations include alterations in temperature, nutrients, and sedimentation, which can have significant consequences for the aquatic ecosystem [22]; furthermore, tropical regions often support higher levels of beta diversity and endemic biodiversity, which are diminished with the loss of small alpine glaciers [22, 23].

While both tropical and temperate biomes face threats from glacier retreat, it is the tropical high-elevation specialist species that are particularly endangered by accelerated deglaciation. These species are adapted to the unique environmental conditions provided by tropical glaciers and may struggle to survive in the absence of these habitats [21–23]. A meta-analysis conducted on a global synthesis explains that glaciers have a negative effect on natural populations and communities living in fjords, freshwaters, and forefields, as shown in **Figure 4**, suggesting that most

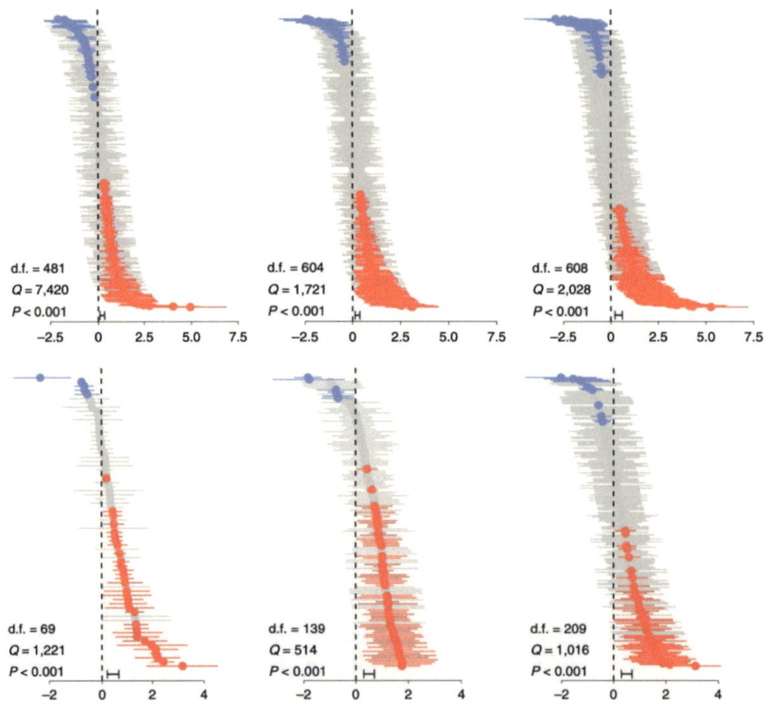

Figure 4.
Population and community responses to the effects of glaciers. Global results obtained from fjords, freshwaters, and forefields. Effect sizes and 95% confidence intervals for taxon abundance (population; top graphs) and diversity (community; bottom graphs) responses to the influence of glaciers in fjord, freshwater, and forefield systems. Colors indicate significantly negative (blue), positive (red), or non-significant (gray) effect sizes corresponding to positive (blue) and negative (red) effects of the glacier on biodiversity [23].

sites are likely to display a gain in local species abundance and richness as glaciers retreat, a pattern frequently reported around the globe. However, the local responses of populations and communities varied considerably. Biodiversity response to the influence of the glacier was independent of the latitude or altitude of the study site, suggesting that environmental filtering linked to glacial influence dominates over regional factors; although regional glacier area, melting rate and site isolation were significant predictors of the biodiversity response, and they were not as strong as filtering processes. [23].

To improve our understanding of the ecological changes following complete glacier disappearance, comparative studies across global mountain ranges are essential, and such studies can help us identify and predict the patterns of species colonization, community dynamics, and alterations in hydrological regimes that occur as a result of glacier retreat [9, 24, 25]. Additionally, comprehensive research approaches and monitoring techniques, such as the use of unmanned aerial vehicles (UAVs) for high-resolution monitoring of tropical glacier change, are necessary to capture their unique characteristics and dynamics [26]. As they exhibit distinct ecological effects compared to temperate regions, there are differences in deglaciation rates, environmental conditions, and species responses, although we still need research approaches and monitoring techniques for understanding and mitigating the impacts generated by future climate change. This can be achieved by comprehensively studying and addressing the ecological implications of glacier melting in the tropics, and with this input, we can work toward effective strategies, policy formulation, and cross-sectorial cooperation, in order to safeguard these fragile ecosystems in the face of climate change.

2.2 Ecological dynamics of highland glacier-influenced streams

Glacier-fed streams in highland areas are characterized by unique ecological dynamics due to the influence of glacial meltwater. These streams are subject to spatial and temporal variations in environmental conditions which in turn affect the diversity and distribution of benthic and hyporheic fauna [1]. Glacier retreat is a global phenomenon that has significant impacts on the hydrological cycle, downstream ecosystem structure, and functioning [16]. Therefore, glaciered environments are highly vulnerable to climate change with intricate connections between atmospheric forcing, snow packs/glacier mass-balance, stream flow, water quality, hydrogeomorphology, and river ecology [27], just as the ecological dynamics of highland glacier-influenced streams are complex and influenced by various factors, including flow reduction, temperature, community structure, nutrient concentrations, and species adaptations [3, 16, 28–30]. Taxonomic richness increases from upstream to downstream for various organisms including cyanobacteria, diatoms, invertebrates, and vertebrates [30]. Although implications on future altitudinal distribution of highland species and their interaction with glacier adapted species are unknown, understanding these dynamics is crucial for assessing the impact of climate change on these fragile ecosystems and their dwellers.

Concerning environment and biodiversity, glacial surfaces, recently deglaciated terrains and glacial streams, support diverse ecosystems known as "glacial biodiversity," insects particularly ground beetles (carabids) and non-biting midges (chironomids) are the best adapted animals to colonize these habitats; these insects have developed adaptation strategies to cope with extreme cold and anthropogenic heat playing important roles in glacial-ecosystem functioning [31]. Glacier-fed streams serve as important early indicator systems for understanding hydrological and ecological responses to climate change. These aquatic bodies exhibit characteristic flow patterns, water temperature regimes, sediment fluxes, and channel forms. The relationship between aquatic macroinvertebrates taxon richness and latitude/altitude follows a hump-shaped pattern peaking at mid-latitudes [27]. Even in catchments that have not been glaciated for more than a century, specialized cold-water invertebrate communities adapted to meltwater streams persist, serving as critical refugia for mountain biodiversity [6]. Studies conducted in different regions of the world have shed light on the factors shaping the communities in glacier-fed streams, for edxample, in the Italian Alps, the habitat type (benthic vs. hyporheic) was found to be the main determinant of community structure, with the hyporheos exhibiting higher species richness than the benthos. The highly disturbed nature of glacial systems compared to stable spring systems contributes to significant differences in benthic communities between glacier-fed and spring-fed streams. Spatial connectivity is crucial for invertebrate dispersal in the face of climate change highlighting the trophic-sink effect between benthos and hyporheos [32]. In the Patagonia Mountains, glacier-fed streams exhibited unexpectedly high richness in benthic taxa, including endemic elements. Integrating benthic biota and environmental variables revealed that a higher environmental heterogeneity related with spatial dimension (unshaded/shaded reaches, wetland reaches), local resources (detritus, bryophytes), and temperature probably explained the unexpected high richness in benthic assemblages. Upholding that isolated, small glacier-fed streams typical of the Patagonian landscape are highly vulnerable to global warming, as endemic elements could disappear at upper segments being replaced by other species common at rhithral environments. Which might increase local diversity (alfa diversity) but decrease regional (gamma diversity), thus stream functioning can result altered from an ecosystem perspective [33].

A synthesis research across Europe revealed that water temperature and channel stability are the primary drivers of macroinvertebrate community structure in glacier-fed rivers. A conceptual model based on this study predicts the occurrence of macroinvertebrate families and subfamilies based on water temperature and channel stability during the summer meltwater period. The model indicated that glacial rivers support higher macroinvertebrate abundance and diversity during other times of the year, when environmental conditions improve toward a more stable system [34]. The longitudinal distribution of macroinvertebrates in an Ecuadorian glacier-fed stream generally follows the temperate model of decreasing density and increasing richness downstream, although certain chironomid subfamilies show different patterns [2]. Temporal variability in the same type of streams is influenced by daily glacial melt cycles, resulting in unseasonal fluctuations in community composition, with the extent of variability differing among sites [14]. **Figure 5** describes an assessment along the slope of a glacier-fed stream in Ecuador revealing considerable variability in assemblage composition and environmental relationships, despite longitudinal patterns of increasing density, richness, and temperature/stability that align with the temperate model [21]. Consistent with previous findings, spatial temperature/stability dynamics is shown to drive local biological interactions, whilst temporal spatial temperature/stability dynamics also represents similar effects. In an unseasonal tropical glacier-fed stream, flow reduction induced significant changes in benthic fauna community composition within a short period of time, and both algal and herbivore biomass increased in response to flow reduction together with an increase of channel stability showing much less temporal variability in algal biomass, thus supporting the

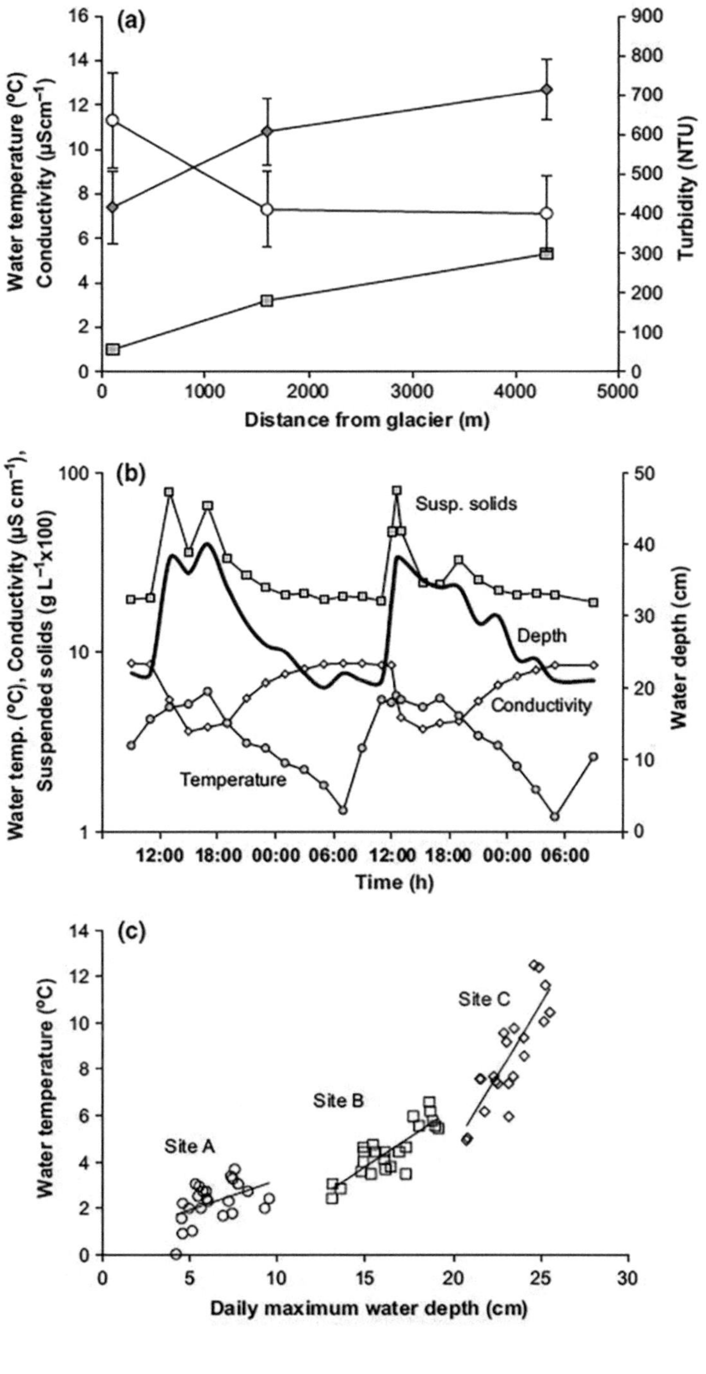

Figure 5.
Spatiotemporal variability in environmental parameters in the Ecuadorian, glacier-fed river. Crespo stream, along the mountain slope at three points; being site a the closest, site B intermediate and site C furthest from the glacier. (a) Mean water temperature, turbidity, and conductivity at the three study sites versus distance from the glacier. (b) Water depth (thick line), suspended solids (filled squares), conductivity (open diamonds), and water temperature (filled circles) measured approximately every 2 h during a 48-h period at site B. Note the logarithmic y-axis. (c) Water temperature versus water depth at peak afternoon flow at each of the three study sites during April 2008. Linear regressions are site a, y = 0.261x + 0.612, R2 = 0.210, P = 0.024; site B, y = 0.502x + 3.731, R2 = 0.710, P < 0.001; and site C, y = 1.243x – 20.210, R2 = 0.704, P < 0.001 [21].

ecological effects of flow and stability changes caused by glacier retreat on aquatic systems [1, 16]. Additionally, a multi-stream survey showed that algal and herbivore biomass abruptly increased below 11% glacier cover in the catchment, further supporting the influence of glacier retreat on stream biota as daily glacial flood pulses in Ecuadorian glacier-fed streams affect macroinvertebrate communities with hydraulic stress; increasing during floods while low-stress areas persist, flood pulses might as well influence local primary production and distribution of herbivores along the stream [16]. Unlike temperate systems, taxon richness and abundance are not significantly impacted, and rare taxa are found in highly stressed habitats [12]. Corroborating with these studies, glacier shrinkage under a warmer climate has profound effects on highland aquatic ecosystems like changes in hydrological regimes, physico-chemical habitats and biota, driven by reductions in sediment load, warmer water temperatures, and increased channel stability. Some taxa may as well benefit from these changes toward stability, while on the other hand, specialized cold-water invertebrate communities including endemic macroinvertebrates may face the risk of extinction resulting in a reduction in regional diversity [35].

Tropical glacier-fed streams ecological dynamics have been studied for some time now, but we still lack relevant ecological information, to reach an accurate assessment of their functioning and the future impacts, such as the glaciers future diminishing or vanishing all over the tropical regions. Even though, we still have acquired some novel information on these complex tropical ecosystems, glacierized catchments in the tropical Andes exhibit complex patterns of species diversity in aquatic macroinvertebrate communities due to the environmental heterogeneity resulting from different water sources [36]. Glacier retreat creates ice-free areas that provide opportunities for studying species colonization patterns and community dynamics, recent advancements in techniques such as environmental DNA metabarcoding and functional trait analysis have expanded our understanding of these processes [9]. Highland glacier-influenced streams are complex ecosystems influenced by multiple environmental factors such as glacier retreat, flow reduction, temperature, and nutrient concentrations. Glacier retreat in the tropics leads to changes in community composition and increased algal and herbivore biomass. Invertebrates in these streams have unique adaptations, and glacial biodiversity includes various organisms. Additionally, benthic and hyporheic fauna distribution affects trophic interactions and dispersal, a topic to asses for similar ecosystems in the tropics. Furthermore, tropical glacier-fed stream communities have complex dynamics affected by daily glacial pulses and spatial heterogeneity, key information for predicting the impacts of glacier loss in tropical regions. So far, we have acquired valuable information about the ecology processes of this complex systems along the last decades, but we still need to reach a deeper understanding about ecosystem's functionality and the real consequences of diminishing glaciers on highland stream; comprehending all these dynamics is essential for assessing the ecological consequences of glacier shrinkage and developing conservation strategies and measures.

2.3 Different stream origins equal distinct diversity scenarios

Streams originating from different sources exhibit distinct environmental conditions resulting in diverse ecological scenarios. We explore the diversity patterns and community compositions of macroinvertebrate taxa in streams sourced from glaciers, glacial-fluvial-limnic pathways, and alpine glacial floodplains. Analyzing research findings from various studies, we will gain insights into the unique traits, distributions, and environmental influences on streams. Aquatic ecosystems in high latitude and altitude environments are strongly influenced by cryospheric and hydrological processes. The current phase of global climate warming has led to the shrinking of glaciers, which will have significant impacts on stream ecosystems [16, 35]. Understanding the biodiversity distributions and compositional variation across hydrologically connected habitats along glacial-fluvial-limnic pathways can provide insights onto the mechanisms underlying glacial community organization and ecosystem processes. A practical example of its relevance, is the spatial patterns of community composition and assembly mechanisms of cyanobacteria, diatoms, invertebrates, and vertebrates along a glacial-fluvial-limnic pathway on the Tibetan Plateau. Where it was found that the taxonomic richness increasing from upstream to downstream for all groups and found significant correlations between community compositions and various environmental factors. Also, it was revealed that stochastic processes played a prominent role in microorganism community assembly, while macroorganisms were influenced more by environmental conditions and spatial factors [30]. It was also Implied that the academy (especially limnologists), decision makers, and stakeholders are facing an intricate situation on mitigating climate change on a complex but must vital ecosystem. Aiming to deliver a solid spatial background on the topic, we will use the information obtained from three neighboring glacier-fed streams in the tropical Andes of Ecuador, where longitudinal patterns in density, taxon richness, composition of macroinvertebrate assemblages, and its driving factors were evaluated. Results showed that density, number of taxa, rarefied richness, and ordination coordinates increased logarithmically with distance from the glacier (**Figure 6**), these faunal metrics were related to altitude, glacier coverage of catchment, conductivity, temperature, and channel stability. The number of taxa varied among sites within the upper and middle reaches of the streams but not among the lower sites, as illustrated in **Figure 7** assemblage composition differed among middle and lower sites, reflecting changes in environmental harshness and its drivers along the streams [37].

Temporality also plays a role on ecosystem dynamics as the successional patterns of aquatic invertebrates, ground beetles, terrestrial plants, and soil eukaryotes in the foreland of the Ecuadorian Carihuairazo glacier was evaluated and described in **Figure 8**. The effects of environmental conditions and the age of deglaciation on community composition were analyzed. The results showed that diversity increased with time since deglaciation, especially among passive dispersers as noted in **Figure 9** a, b, and e literals, suggesting that dispersal played a key role in structuring the succession of the glacier foreland. Spatial beta-diversity was mainly attributed to nestedness and turnover, indicating different mechanisms driving diversity in different taxa [24]. Although, spatial and temporal aspects are not the only driving factors defining community structure, a good example of how different stream features may modulate ecosystem's diversity and functionality is exposed in a study on a glacierized catchment in the tropical Andes. Different stream sites with varying water sources and glacier coverage showed varying levels of diversity and commonness, as environmental variables such as turbidity, temperature, and nutrient concentrations influenced the presence and density of macroinvertebrate taxa. In this case, glacier influence was found to increase biodiversity in glacierized catchments by favoring the presence of rare taxa and taxa turnover [36], delivering better understanding of the colonization dynamics before the glaciers probable vanishing especially for aquatic fauna as their habitat will completely disappear. Knowledge on streams spatial and temporal dynamics, from different sources, reveals diverse diversity scenarios influenced by environmental conditions and community compositions. Equatorial glacier forelands

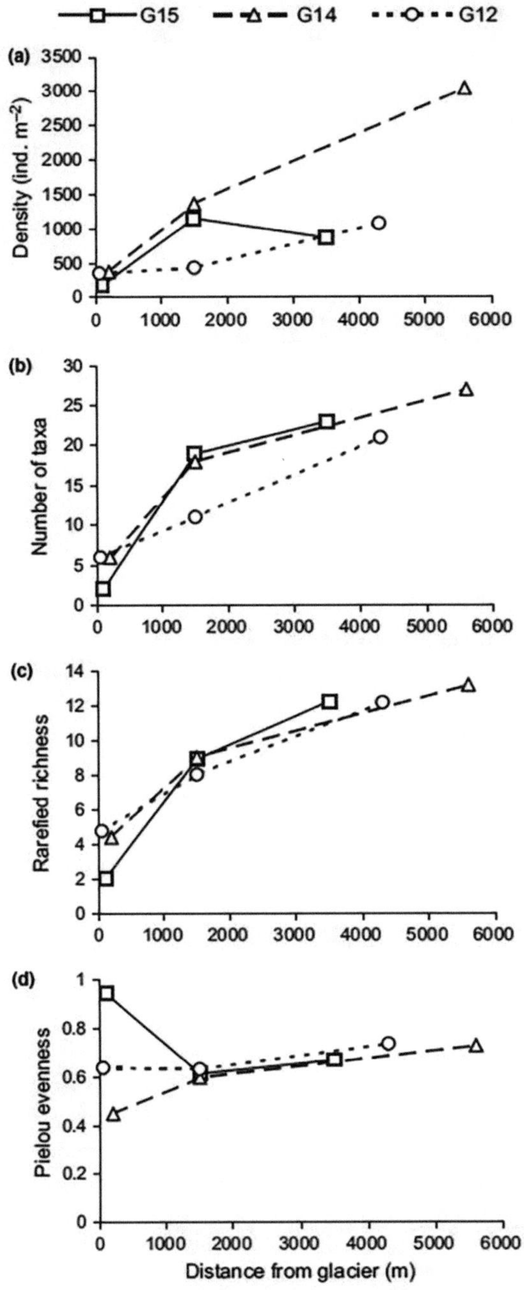

Figure 6.
Macrofaunal longitudinal patterns. Density of individuals (a), number of taxa (b), rarefied richness (c), and Pielou's evenness (d) as a function of distance from the glacier, showing the three study sites along each of the three Antisana streams [37].

and glacier-fed streams also exhibit distinct diversity driven by dispersal processes and environmental factors. Tropical glacier-fed streams support unique macroinvertebrates adapted to daily harsh conditions, and shrinking glaciers will significantly impact future stream biodiversity, requiring further research and conservation efforts

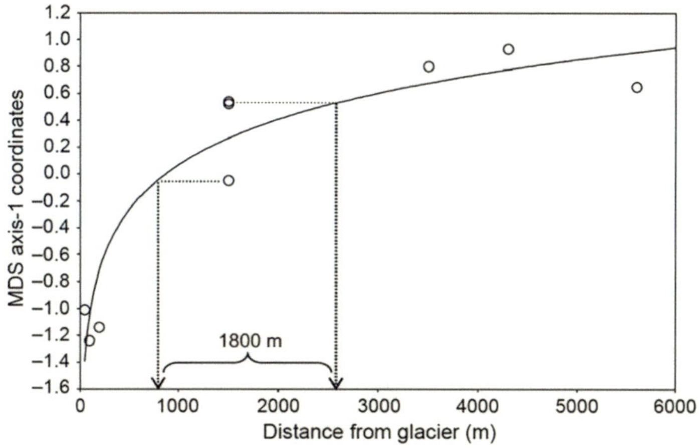

Figure 7.
Taxa abundance variation. Logarithmic fit to the relationship between axis-1 coordinates from a MDS ordination of the nine Antisana stream sites as a function of their longitudinal distance from the glacier (y = 0.4865Ln(x)) 3.293; r2 = 0.89; P = 0.0002). The diagram illustrates that, following the general relationship, the variation among the three replicate sites located at 1500 m from the glaciers corresponds to a longitudinal distance of 1800 m (800–2600 m) [37].

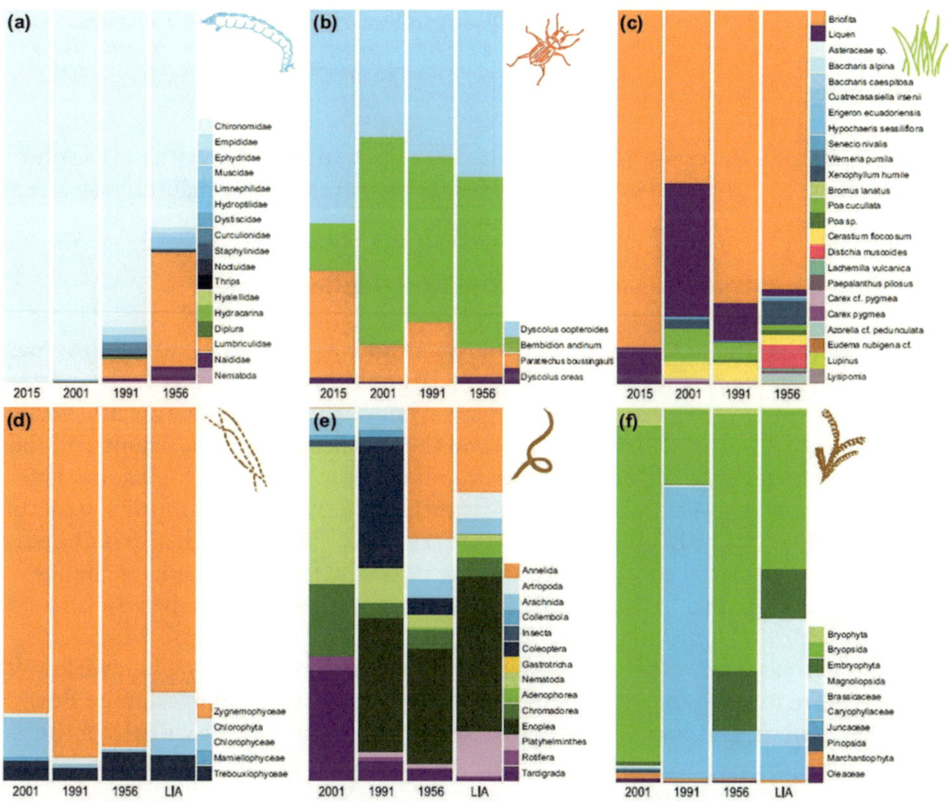

Figure 8.
Dominant taxa relative abundance. Samples collected for (a) aquatic invertebrates, (b) ground beetles, (c) terrestrial plants, (d) soil algae, (e) soil invertebrates, and (f) soil plants along the gradient of deglacierized zones, dating from 2015, 2001, 1991, 1956, and the little ice age (LIA) about 150–250 years ago [24].

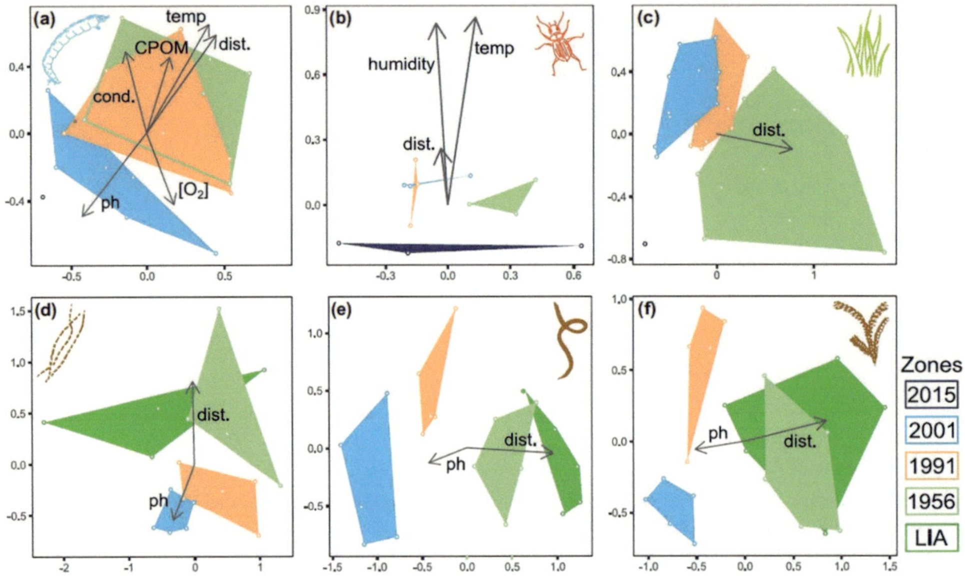

Figure 9.
Non-metric multidimensional scaling ordination. Of (a) aquatic invertebrates, (b) ground beetles, (c) terrestrial plants, (d) soil algae, (e) soil invertebrates, and (f) soil plants along the gradient of deglacierized zones dating from 2015, 2001, 1991, 1956, and LIA. Vectors of all tested environmental variables were overlaid onto the ordination space. The arrow indicates the direction of most rapid change in the variable, and its length is proportional to the correlation between ordination and the variable [24].

because predicting the consequences of glacial retreat on biodiversity needs a solid understanding of this patterns and processes, as they are vital for maintaining stream integrity amidst future environmental changes.

2.4 Environmental patterns and ecosystem functioning

Understanding the biodiversity distributions and compositional variation across hydrologically connected habitats along glacial-fluvial-limnic pathways can provide insights into the mechanisms underlying glacial community organization and ecosystem processes [16]. In an effort to understand the community compositions and the assembly mechanisms of different organisms along alpine streams, the habitat type (benthic or hyporheic) was found to significantly influence the community structure with the hyporheic zone having higher species richness. The research also highlighted the importance of spatial connectivity and the potential impact of climate change on invertebrate dispersal [32], suggesting that more information on the relations between benthic and hyporheic communities is needed as it might affect habitat conditions such as food resources for stream macrofauna. Especially when we know that glaciers are melting due to a warmer climate, there will be changes in the flow of water, sediment, and temperature of these rivers. These changes will affect the composition of organisms in the rivers with an expected increase in the abundance and diversity of microorganisms, algae, macroinvertebrates, and fish; additionally, the reduction of glacial influence may disturb off-channel habitats and lead to the vulnerability of certain species [37].

At the Carihuairazo Mountain, located in the tropical Andes, the impact of different stream features was examined, such as water sources and glacier coverage, on the

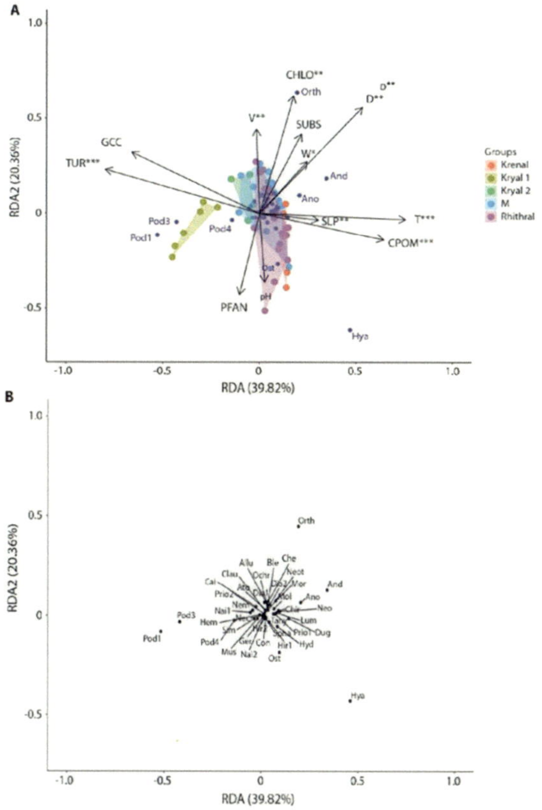

Figure 10.
*Redundancy analysis (RDA). For abiotic, biotic variables and benthic invertebrate communities sampled in 51 study sites, grouped by a) stream site types and b) macroinvertebrate assemblages where each point represents the benthic invertebrate assemblage of 5 pooled Surber samples from each of the 51 study sites. Macroinvertebrate densities (Ind m-2) were Hellinger transformed and environmental variables were log (x + 1) transformed. Significant codes: *** P < 0.001; ** P < 0.01 and * P < 0.05 obtained by 999 permutations. And: Andesiops sp., Ble: Blepharicera sp., Ger: Geranomyia sp., hem: Hemerodromia sp., Hir1 &2: Hirudinea type 1 and 2, Hya: Hyalella sp., Mor: Mortoniella sp., Nec: Nectopsyche sp., Orth: Orthocladiinae, Ost: Ostracoda, Pod1, 3 & 4: Podonominae type 1, 3, and 4, Spha: Sphaeriidae. TUR: Turbidity, GCC: Glacier cover per catchment, PFAN: Pfankuch index, V: Current velocity, CHLO: Chlorophyll a, SUBS: Substrate types, W: Stream width, D: Stream depth, SLP: Stream slope, CPOM: Coarse particulate organic matter, and T: Water temperature [36].*

diversity and functionality of ecosystems. It is found that environmental variables influenced the presence and density of certain organisms, glaciers increased biodiversity by promoting the presence of rare taxa. Moreover, it is revealed that diversity increased over time since deglaciation, with dispersal playing a key role in the succession of the glacier foreland. **Figure 10** explains how density, taxon richness, and composition of macroinvertebrate assemblages were found to be influenced by several factors such as altitude, glacier coverage, conductivity, temperature, and channel stability. The succession of various organisms was examined, and the impact of environmental conditions and deglaciation age on community composition was analyzed. The results indicated that diversity increased over time, particularly among passive dispersers, suggesting that dispersal played a crucial role in shaping the succession of the glacier foreland and similar environments. Spatial beta-diversity was influenced by nestedness and turnover, indicating distinct mechanisms driving diversity in different taxa [36].

Studies conducted on streams originating from different sources have revealed that these streams exhibit distinct diversity scenarios, primarily influenced by variations in environmental conditions and community compositions. For instance, glacier-fed streams support macroinvertebrate communities that have adapted to harsh conditions prevalent in these environments, while on the other hand, glacial-fluvial-limnic pathways exhibit compositional variations influenced by other environmental factors and assembly mechanisms. Similarly, alpine glacial floodplains highlight the significance of benthic and hyporheic connectivity in shaping invertebrate dynamics [16, 32], topic that as mentioned before must get more attention in the future. Equatorial glacier forelands and glacier-fed streams also display distinct diversity scenarios based on their different origins. In the case of glacier forelands, the succession of these areas is primarily influenced by dispersal processes, while composition and richness of macroinvertebrate assemblages in glacier-fed streams are shaped by environmental factors such as temperature, stability, and glacier coverage [11, 24, 36]; consistently with the evidence presented in the previous section (Section 2.3), based on the information obtained so far, future biodiversity of streams in cold environments will be significantly impacted by the shrinkage of glaciers and the resulting changes in runoff, water source contributions, and physico-chemical habitat. To fully grasp the hydroecological response of river systems to shrinking glaciers and ascertain the implications for biodiversity, conducting long-term ecological studies and implementing conservation efforts are essential for the protection and management of these vulnerable freshwater ecosystems.

2.5 Life traits of a multi-environmental riverine system

Tropical highland glacier-fed streams represent a multi-environmental riverine system composed by diverse habitats and ecological conditions influenced by factors such as glacial meltwater, groundwater, rainfall, and climate change. These streams are characterized by the influence of glacial meltwater resulting in distinct physico-chemical properties and specialized communities of organisms; thus, understanding the life traits of organisms inhabiting these systems is crucial for comprehending their adaptations, functional strategies, and responses to environmental changes. Nevertheless, the lack of knowledge about their traits makes it difficult to predict the impacts of global warming and glacier melt on these ecosystems [5, 11, 13, 19, 24, 36]. Here, we explore the life traits of a multi-environmental riverine system, with a particular focus on glacier-fed streams and their ecological dynamics, and information about the structural and functional aspects of these unique and complex ecosystems is presented in this section of the chapter.

The retreat of glaciers caused by climate change has significant effects on river ecosystems particularly in terms of physicochemical changes and impacts on aquatic communities [36]. Studies in the European Alps have revealed significant connections between declining glacier cover, temperature, community structure, diversity, feeding strategies, early life development, body mass, and growth of invertebrates [5]. The impact of environmental conditions and the age of deglacierization on colonization and successional patterns in glacier forelands has also been assessed. Passive dispersal plays a significant role in increasing diversity over time with pioneer communities being dominated by species with flexible feeding strategies and high dispersal abilities. In addition, glacier-fed rivers provide extreme environments for studying food web dynamics characterized by low taxon richness, highly connected individuals, and short mean food chain length as the dominant macroinvertebrates in these rivers exhibit omnivorous and detritivorous feeding strategies, aligning with general food web scaling predictions [13].

A global study has demonstrated consistent responses of community trait composition and diversity to reduced glacier cover across different biogeographic regions; using data from 363 records comprising over 1.23 million invertebrates collected from rivers across nine biogeographic regions on three continents. The study focused on the responses of community trait composition and diversity to replicated gradients of reduced glacier cover. Finding that decreasing glacier cover is locally associated with an increase in functional diversity primarily driven by dispersal limitation and environmental filtering [17], although as **Figure 11** depicts, overall glacier-influenced highland hydrological ecosystem functional diversity is expected to decrease due to the disappearance of intermediate disturbances, such as flood pulses and dramatic thermal and conductivity fluctuations, among other associated variables. Researchers show that glacier influence drives instream food source distribution, consequently meddling with faunal distribution according to the necessities of each species functional feeding group, sieving more generalist species toward harsher environments caused by glarier influence (GI), and more specialist species where glacier-influenced variables decrease their abrasive effects on particulate organic matter (POM) and perifiton growth (Chl. a). Thus, retreat and eventual diminish of glaciers in the tropical Andes will affect the trophic structure and ecosystem processes in high Andean meltwater and other streams sharing the same system. **Figure 12** shows us how the usual global pattern repeats in the Ecuadorian Andes as macrofaunal abundance decreases with increasing glacial influence; but in contrast, taxon richness and the number of functional feeding groups followed a hump-shaped relationship (**Figure 12A**). This interesting dynamic appears due to opportunistic and generalist feeding modes, preferring areas with high glacial influence where physical abrasion and microbial decomposition played a major role in detritus decomposition. However,

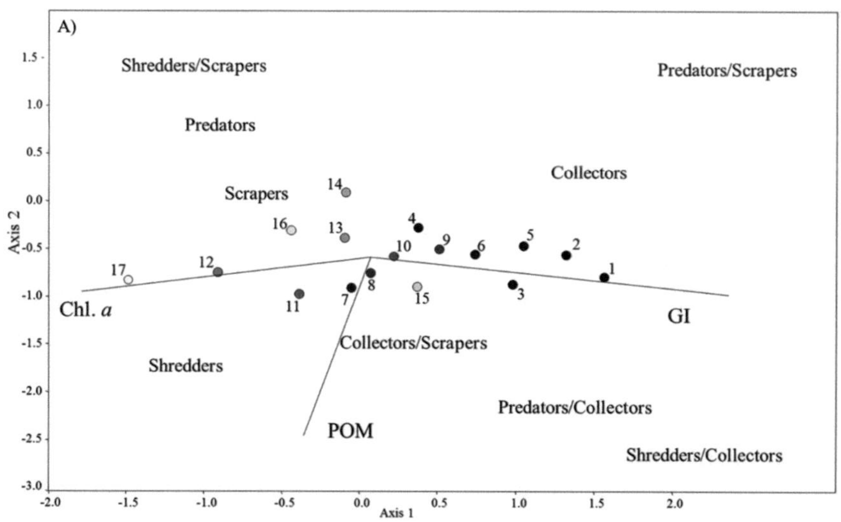

Figure 11.
Canonical correspondence analysis (CCA). Biplot on functional feeding groups, and with the environmental variables (vectors) particulate organic matter (POM), chlorophyll a (Chl. A), and the glaciality index (GI) plotted as correlations with site scores (scaling type 1). B) Density, number of taxa, and number of FFGs of benthic macroinvertebrates along the gradient of glaciality index (GI). Results of exponential (density) and quadratic regressions (taxon richness and number of FFGs) are shown [11].

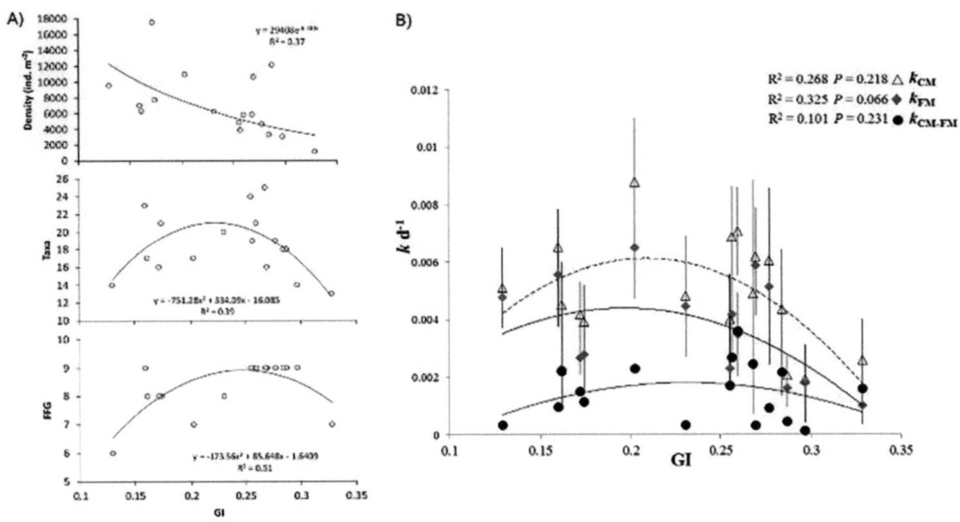

Figure 12.
Ecological taxa and function dynamics. A) Density, number of taxa, and number of FFGs of benthic macroinvertebrates along the gradient of the glaciality index (GI). Results of exponential (density) and quadratic regressions (taxon richness and number of FFGs) are shown. B) Decomposition rates of Calamagrostis sp. in coarse mesh bags (kCM, dashed line), fine mesh bags (kFM), and their difference (kCM – kFM) along the glaciality influence index (GI). Results of the quadratic regressions are shown. [11].

at an intermediate glacier influence gradient, we have not only the most functional feeding groups but also the highest litter decomposition rates (**Figure 12B**), supporting the notion of intermediate disturbance modulating biodiversity and suggesting that glacier loss may partially replace physical abrasion with biological shredding [11].

Glacierized catchments in the tropical Andes exhibit high levels of environmental heterogeneity leading to increased species turnover rates and regional diversity, and higher glaciality is associated with an increase in small-sized taxa that are adapted to harsh environmental conditions. The functional diversity decreases with stronger glacier influence, while the assembly of communities in tropical glacierized catchments is driven by both local and regional processes, with environmental filtering playing a significant role [36]. As the aforementioned global study points out, while functional trait analyses have been used to uncover these mechanisms, their application to invertebrates is relatively underdeveloped compared to other groups of organisms [17].

The life traits of a multi-environmental riverine system are shaped by the unique environmental conditions and ecological processes associated with these ecosystems [5]. As we become more aware of the distinct functional properties of invertebrates in glacier-fed streams and their relationships with declining glacier cover, climate change, and glacier retreat, we are more aware of their influence on the multitrophic effects of mountain rivers, which impose implications for aquatic communities [19]. The acquired knowledge highlights the distinct functionality of invertebrates in glacier-fed streams and their relationships with declining glacier cover [5]. Therefore, understanding the living features of organisms in multi-environmental systems, particularly glacier-fed streams, is crucial for comprehending their adaptations, functional strategies, and responses to environmental changes. By enhancing our knowledge of these systems, we can better predict and mitigate the impacts of climate change and glacier melt on river ecosystems and protect vulnerable species.

2.5.1 Future impacts: melting and diminishing glaciers

The impacts of climate change on glacier-fed river systems have significant implications highland's ecosystem resilience especially on glacier-influenced ecosystems. These impacts are noticeable in various aspects including water resources, biodiversity, and ecosystem functioning; understanding these impacts and their consequences is crucial for addressing the challenges and enhancing ecosystem resilience [5, 11, 15, 19, 33, 36]. Alpine streams are known for their high species and genetic diversity which is a result of habitat insularity and environmental heterogeneity, amd climate change is expected to impact biodiversity in alpine streams across different levels from microorganisms to communities [38]. Ecosystem functioning is also expected to be altered as glacier-fed systems are characterized by distinct abiotic conditions and ecological processes [39]. These changes influence the growth of primary producers and the composition of aquatic communities impacting the overall functioning of river ecosystems [13]. The altered environmental conditions and loss of glaciers threaten ecosystem services provided by these systems, such as water supply, hydrological regulation, and tourism [15, 40]. The Hindu Kush Himalaya Mountains are a vital source of water for a large population in Asia. However, climate change is causing accelerated melting of snow and glaciers in these mountains, posing a threat to water security for millions of people [5]. The reduction in glacial meltwater contribution to river flow affects the availability of water resources for sectors such as agriculture, hydropower, and cities [5, 19]. Similarly, important glaciers in the inner tropics, such as the "Conejeras" Glacier in Colombia, are rapidly retreating due to warming temperatures causing concern on local communities.

The longitudinal distribution of benthic macroinvertebrate assemblages in glacier-fed streams is influenced by glacial run-off patterns [21]. Short-term variations in glacial run-off reflected, and changes of temperature, conductivity, and turbidity can result in comparable environmental conditions occurring at different distances along the stream. These findings suggest that glacial retreat can lead to shifts in altitudinal and longitudinal species ranges potentially impacting overall ecosystem function and resilience [11, 21]. Permanent changes in run-off caused by glacial shrinkage may result in multidirectional shifts in species ranges including a potential downward shift. Specialist species adapted to glacial conditions may face increased vulnerability, while generalist taxa may have opportunities to expand their ranges. These changes have implications for the functioning and stability of glacier-fed ecosystems and their ability to withstand future environmental challenges and the ecological resilience to return to its former state [11]. Glacier-fed river systems support unique and specialized aquatic communities, although the rapid shrinking of glaciers due to climate change poses a significant threat to freshwater biodiversity. The disappearance of glaciers leads to a reduction in local and regional species diversity including endemic species, as the loss of glacial meltwater impacts macroinvertebrate communities and disrupts the ecological balance of river ecosystems [15]. Furthermore, the warming of glacial streams alters the community structure of primary producers such as epilithic algae with potential negative consequences for ecosystem functioning [38]. The hydrological implications of glacier retreat are significant as glaciers serve as freshwater reservoirs. The melting flows increase, but in the long-term, a decrease in water resources is expected as glaciers become smaller [7]. Changes in river runoff and seasonality of flow are projected to increase in the future, further impacting water supplies globally [15, 40].

As mentioned before, the impacts of climate change on glacier-fed river systems have far-reaching consequences on overall ecosystem resilience, and understanding and addressing these impacts through global cooperation, adaptation measures, and research is crucial for maintaining the future integrity and functionality of these complex ecosystems [5, 15, 19, 33, 38, 39]. The cumulative effects of these impacts on water resources, biodiversity, and ecosystem functioning will affect the ecosystem's resilience of glacier-fed and glacier-influenced river systems. To address these challenges and enhance ecosystem resilience, several measures are suggested [5]. For instance:

- Global emission reduction: mitigating climate change by reducing global greenhouse gas emissions is crucial to slow down the rate of glacier melting and preserve the integrity of these ecosystems.

- Regional cooperation: as collaboration among countries sharing the glacier-influenced highland river systems is essential to develop coordinated adaptation strategies and manage shared water resources effectively.

- Adaptation support: for underdeveloped countries, most of them located in the tropics, increasing technical and financial support for adaptation efforts is necessary to assist communities in coping with the changing conditions and developing sustainable practices.

- Interdisciplinary science: robust and interdisciplinary scientific research is needed to understand the complex interactions between climate change, glacier retreat, and ecosystem responses.

This knowledge can inform policy decisions and guide effective management strategies [5, 15, 19, 40].

Glacier shrinkage is a clear indication of global warming and climate change. Mountain glaciers have been retreating worldwide during the twentieth century. This shrinkage has significant implications for water resources due to changes in river hydrology and morphology caused by glacier loss. Alterations are projected to have the greatest impact compared to other hydrological systems. These changes affect riverine and near-shore marine environments, biodiversity, and ecosystem services provided by glacier-fed rivers including water supply for agriculture, hydropower, and consumption [26], suggesting that glacial retreat can lead to shifts in altitudinal and longitudinal species ranges, potentially affecting biodiversity and ecosystem resilience [21]. Permanent changes in runoff caused by glacial shrinkage may result in multidirectional shifts in species ranges including potential downward shift. Specialist species adapted to glacial conditions may face increased vulnerability, while generalist taxa may have opportunities to expand their ranges. These changes have implications for the functioning and stability of glacier-fed ecosystems and their ability to withstand future environmental challenges, which also affects the water local and regional water supply. Thus, if this species-functionality relationship disappears, water quality will be affected, with repercussions on the human populations that are supplied by this vital resource; yet a comprehensive framework for predicting biodiversity responses to glacier retreat globally is still lacking.

2.5.2 Future impacts: downstream implications

Water supply nowadays is already scarce at several populated regions of the globe. Our demography keeps increasing worldwide, and we still lack strategies to handle such a challenging problem [41–49]. Several glaciers, such as Ecuador's Cotacachi, have already disappeared, providing an early glimpse of upcoming consequences, revealed by Vergara 2007. The area around Cotacachi is experiencing declines in agriculture, tourism, loss of biodiversity, waterless streams, and a decrease in water levels, which already have led to more water conflicts. These are expected to worsen and more are expected over time [50]. A systematic review of conducted research on human impacts of glacier meltwater variability in mountain ranges worldwide identified the main areas of existing research that could more accurately detect and attribute glacier runoff and human impacts, grapple with complex and intersecting spatial and temporal scales, and implement transdisciplinary research approaches to study glacier runoff to redefine the glacier-water problem around human societies rather than simply around ice and climate, by systematically evaluating human impacts in different mountain regions [46]. This perspective has been gradually used by several scientists over the last decades with positive results, increasing our understanding on how regional and local freshwater hydrological ecosystems will be affected by climate change and its repercussion over our future society. Fortunately, as indicated previously in this section, we are already aware of the matter, and nowadays, there are several initiatives on gathering information and in some cases designing strategies to keep water resource's quantity and quality. As a local example, the municipality of Quito (Ecuador) adopted Quito's Climate Change Strategy (QCCS) in 2009, on which, in coordination with key stakeholders, including the academia, the municipality is currently implementing a series of adaptation and mitigation measures in key sectors in response to current and projected impacts [51]. Such as this, several efforts are globally taking place on gathering knowledge about ecological and social responses implications of glacier shrinking and eventual disappearance of tropical glaciers [45–48, 52–55].

Glacier contributions to river discharge are not well known on a regional or global basis, nor are the populations at risk to future glacier changes. When estimating the upper bounds on the fraction of river discharge attributable to glacier, globally 370 (140) million people live in river basins where glacier sources contribute at least 10% (25%) of river discharge on a seasonal basis. Most of this population is in the High Asia region [56]. In the Indo-Gangetic Plains, snow and glacier melt modulates the seasonal pattern of river flows together with groundwater. Necessity varies strongly in space and time and is highest in the Indus basin, where in the pre-monsoon season, up to 60% of the total irrigation withdrawals originate from mountain snow and glacier melt; although dependence in the floodplains of the Ganges is comparatively lower, meltwater is still essential to provide enough water to sustain 38 million people during the dry season, and climate change is expected to weaken this modulating effect [57]. Similar results were found in Pakistan at the Burgay watershed, where glaciers/snow water dependence was evaluated, revealing that during winter, flow discharge was relatively low, while during summer, major flow was generated by snow/glacier melt due to the increase of glacier melt along with rising air temperature; snow/glacier melt was found to be the major contributor (79%) during the summer months, while spring water contribution was 21%. However, in winter months, only spring water is available for drinking and domestic use and hardly any was available to meet irrigation needs; thus, agrarian economy of the Burgay watershed is highly dependent on glacier/snow, and farmers face severe water shortage during winter [58]. In contrast, the application of coupled modeling with regional climate models for the central Himalayan Lhasa River basin was obtained. Although glaciers have retreated and will continue to retreat, water availability is and will be primarily determined by monsoon precipitation and snowmelt; ice melt from glaciers is and will be a minor runoff component in summer monsoon-dominated Himalayan river basins [52]. The lesser sustainability of glaciers in the Himalayas apparently will be compensated by monsoon and change in rainfall pattern as long as water can be stored or accumulated in the quantities needed to overcome the dry season [59]. In the Himalayas, different responses to water scarcity are needed as it depends on local conditions and should include the construction of new irrigation channels, installation of pipes construct water storages, lifting of ground water or river, and even building of artificial ice reservoirs to meet irrigation requirements [55].

On the other hand, almost all of the world's glaciers in the tropical latitudes are located in the Central Andes (Peru, Bolivia, Ecuador, and Colombia). Due to their high altitude, to the high level of radiation, and to the tropical climate dynamics, they all are particularly threatened by climate change, as a result of not only warming but also of changing variability of precipitation. Many glaciers are of crucial importance for the livelihood of the local populations and even for three capitals, Lima (Peru), La Paz (Bolivia), and Quito (Ecuador), which depend on them for water and energy supplies [60]. This accelerated melting of glaciers is expected to have a negative effect on the water resources of mountain regions and their adjacent lowlands, with tropical mountain regions being among the most vulnerable. In order to quantify those impacts, it is necessary to understand the changing dynamics of glacial melting, but also to map how glacial meltwater contributes to current and future water use which often occurs at considerable distance downstream of the glacier. While the dynamics of tropical glacial melt are increasingly well understood and documented, major uncertainty remains on how the contribution of tropical glacial meltwater propagates through the hydrological system and hence how it contributes to various types of human water use in downstream regions [45, 61]. In the tropical Andes, the trend has been quite negative over the past 50 years with a mean mass balance deficit for glaciers that is slightly more negative than the one computed on a global scale. Monthly mass balance measurements performed in Bolivia, Ecuador, and Colombia show that variability of the surface temperature of the Pacific Ocean is the main factor governing variability of the mass balance at the decadal timescale, while precipitation did not display a significant trend in the tropical Andes and consequently cannot explain the glacier recession [53]. Evidencing that tropical glacier melt is mainly driven not by local, regional and global climatic dynamics will affect local hydric systems. Research at the Ecuadorian Andes shows how glacier cover contributes to downstream catchment. As the diurnal variation power of glacier-fed streams decreased downstream with the addition of non-glacial tributaries, the diurnal variation power and the percentage of the glacier cover in the catchment were significantly positively correlated [62].

Corresponding to the experiences from Asia [52, 55, 57–59], at the tropical Andes regional mapping of water demand in downstream regions of the major tropical glaciers combined with regional water balance model aiming to determine the dominant spatiotemporal patterns of the contribution of glacial meltwater to human water use, results showed that the number of users relying continuously on water

DOI: http://dx.doi.org/10.5772/intechopen.1005016

resources with a high long-term average contribution from glacial melt is low, but this reliance increases sharply during drought conditions; a large proportion of domestic and agricultural users are located in rural regions where climate adaptation capacity tends to be low [61]. Likewise, the supply of glacier water to La Paz city, Bolivia, between 1963 and 2006 was assessed at annual and seasonal timescales based on the mass-balance quantification of 70 glaciers located within the drainage basins of La Paz. Glaciers contributed 15% of water resources to the city at an annual scale (14% in the wet season, 27% in the dry season) [54]. In contrast to the experience in Burgay [58], this research found that despite the loss of 50% of the glacierized area during the study period runoff at La Paz did not change significantly, showing that increase in ice melt rates compensated for reduction in the surface area of the glaciers; in the future, assuming complete disappearance of the glaciers and no change in precipitation, runoff should diminish by 12% at an annual scale, 9% during the wet season, and 24% during the dry season [54]. Is important to note that this calculation do not involve precipitation changes which are also very likely to occur in the future, and the notion that glacier melt occurs daily over the year in the tropics, without a winter season as in Burgay.

Therefore, the general perception is to investigate more complex regional and local environmental patterns, including spatiotemporal shifts in downstream watershed dynamics, for the purpose of developing adaptation strategies that should also focus on increasing the natural and artificial water storage and regulation capacity to bridge local dry periods [57, 61, 62]. Thus, based on the experience of present studies and from a technical point of view, we still bear uncertainties to be understood. We are as well brightening the path to more and more complex questions about our future and how to adapt to climate change as a society. Climate change effects on glacier-influenced water resources for both temperate and tropical systems represents an intricate issue that needs to be assessed both regionally and locally, as highlighted by most of the research presented in this section. Accordingly with any science, the acquired knowledge generates at the same time more questions and uncertainties in front of us, bringing to notice our lack of understanding as well as our efforts to comprehend climate change implications, as worldwide academy continues to match this complex challenge.

Author details

Patricio Andino Guarderas[1*] and Rodrigo Espinosa Barrera[2]

1 Independent Researcher and Environmental Consultor, Quito, Ecuador

2 Faculty of Life Sciences, Amazon Regional University Ikiam, Tena, Ecuador

*Address all correspondence to: pj.andino.guarderas@gmail.com

References

[1] Jacobsen D, Dangles O. Ecology of High Altitude Waters. Oxford; online edn.: Oxford Academic; 2017. DOI: 10.1093/oso/9780198736868.001.00012

[2] Jacobsen D, Dangles O, Andino Guarderas P, Espinosa R, Hamerlík L, Cadier E. Longitudinal zonation of macroinvertebrates in an Ecuadorian glacier-fed stream: Do tropical glacial systems fit the temperate model? Freshwater Biology. 2010;**55**:1234-1248. DOI: 10.1111/j.1365-2427.2009.02348.x.3

[3] Vanderwall J, Muhlfeld C, Tappenbeck T, Giersch J, Ren Z, Elser J. Mountain glaciers influence biogeochemical and ecological characteristics of high-elevation lakes across the northern Rocky Mountains, USA. Limnology and Oceanography. 2023;**69**:37-52. DOI: 10.1002/lno.12434

[4] Wigmore O, Bryan M. Monitoring tropical debris covered glacier dynamics from high resolution unmanned aerial vehicle photogrammetry, cordillera Blanca, Peru. The Cryosphere Discussions. 2017;**11**:1-27. DOI: 10.5194/tc-2017-31

[5] Azam MF, Kargel J, Shea J, Nepal S, Haritashya U, Srivastava S, et al. Glaciohydrology of the Himalaya-Karakoram. Science. 2021;**373**:6557. DOI: 10.1126/science.abf3668

[6] Rangecroft S, Harrison S, Anderson K. Rock glaciers as water Stores in the Bolivian Andes: An assessment of their hydrological importance. Arctic Antarctic and Alpine Research. 2015; 47, 89-98. DIO: 10.1657/AAAR0014-029.

[7] Morán-Tejeda E, Ceballos J, Peña K, Lorenzo-Lacruz J, López-Moreno J. Recent evolution and associated hydrological dynamics of a vanishing tropical Andean glacier: Glaciar de Conejeras, Colombia. Hydrology and Earth System Sciences. 2018;**22**:5445-5461. DOI: 10.5194/hess-22-5445-2018

[8] Francou B, Coudrain A. Glacier shrinkage and water resources in the Andes. Eos, Transactions American Geophysical Union. 2005;**86**:415. DOI: 10.1029/2005EO430005

[9] Michoud G, Kohler T, Ezzat L, Peter H, Nattabi J, Nalwanga R, et al. The dark side of the moon: First insights into the microbiome structure and function of one of the last glacier-fed streams in Africa. Royal Society Open Science. 2023;**10**(8):230329. DOI: 10.10.1098/rsos.230329

[10] Ficetola Gentile F, Marta S, Guerrieri A, Gobbi M, Ambrosini R, Fontaneto D, et al. Dynamics of ecological communities following current retreat of glaciers. Annual Review of Ecology Evolution and Systematics. 2021;**52**:405-426. DOI: 10.1146/annurev-ecolsys-010521-040017

[11] Andino Guarderas P, Espinosa R, Crespo-Pérez V, Cauvy-Fraunié S, Dangles O, Jacobsen D. Functional feeding groups of macrofauna and detritus decomposition along a gradient of glacial meltwater influence in tropical high-Andean streams. Water. 2021;**13**:3303. DOI: 10.3390/w13223303

[12] Cauvy-Fraunié S, Andino Guarderas P, Espinosa R, Calvez R, Anthelme F, Jacobsen D, et al. Glacial flood pulse effects on benthic fauna in equatorial high- Andean streams. Hydrological Processes. 2013;**28**:3008-3017. DOI: 10.1002/hyp.9866

[13] Clitherow L, Carrivick J, Brown L. Food web structure in a harsh glacier-fed river. PLoS One. 2013;**8**:e60899. DOI: 10.1371/journal.pone.0060899

[14] Jacobsen D, Andino Guarderas P, Calvez R, Cauvy-Fraunié S, Espinosa R, Dangles O. Temporal variability in discharge and benthic macroinvertebrate assemblages in a tropical glacier-fed stream. Freshwater Science. 2014;**33**:32-45. DOI: 10.1086/674745

[15] Jacobsen D, Milner A, Brown L, Dangles O. Biodiversity under threat in glacier-fed river systems. Nature Climate Change. 2012;**2**:361-364. DOI: 10.1038/nclimate1435

[16] Cauvy-Fraunié S, Andino Guarderas P, Espinosa R, Calvez R, Jacobsen D, Dangles O. Ecological responses to experimental glacier-runoff reduction in alpine rivers. Nature Communications. 2016;**7**:12025. DOI: 10.1038/ncomms12025

[17] Brown L, Khamis K, Wilkes M, Blaen P, Brittain J, Carrivick J, et al. Functional diversity and community assembly of river invertebrates show globally consistent responses to decreasing glacier cover. Nature Ecology and Evolution. 2017;**2**:325-333. DOI: 10.1038/s41559-017-0426-x

[18] Losapio G, Cerabolini B, Maffioletti C, Tampucci D, Gobbi M, Caccianiga M. The consequences of glacier retreat are uneven between plant species. Frontiers in Ecology and Evolution. 2021;**8**:616562. DOI: 10.3389/fevo.2020.616562

[19] Fell S, Carrivick J, Brown L. Multi-trophic effects of climate change and glacier retreat in mountain rivers. Bioscience. 2017;**67**:897-911. DOI: 10.1093/biosci/bix107

[20] Lencioni V. Glacial influence and stream macroinvertebrate biodiversity under climate change: Lessons from the Southern Alps. Science of The Total Environment. 2018;**622-623**:563-575. DOI: 10.1016/j.scitotenv.2017.11.266

[21] Jacobsen D, Cauvy-Fraunié S, Andino Guarderas P, Espinosa R, Cueva D, Dangles O. Runoff and the longitudinal distribution of macroinvertebrates in a glacier-fed stream: Implications for the effects of global warming. Freshwater Biology. 2014;**59**:2038-2050. DOI: 10.1111/fwb.12405

[22] Cuesta F, Llambí L, Huggel C, Drenkhan F, Gosling W, Muriel P, et al. New land in the neotropics: A review of biotic community, ecosystem and landscape transformations in the face of climate and glacier change. Regional Environmental Change. 2019;**19**:1-20. DOI: 10.1007/s10113-019-01499-3

[23] Cauvy-Fraunié S, Dangles O. A global synthesis of biodiversity responses to glacier retreat. Nature Ecology Evolution. 2019;**3**(12):1675-1685. DOI: 10.1038/s41559-019-1042-8. p

[24] Rosero P, Crespo-Pérez V, Espinosa R, Andino Guarderas P, Barragán A, Moret P, et al. Multi-taxa colonisation along the foreland of a vanishing equatorial glacier. Ecography. 2021;**44**:1-12. DOI: 10.1111/ecog.05478

[25] Milner A, Khamis K, Battin T, Brittain J, Barrand N, Füreder L, et al. Glacier shrinkage driving global changes in downstream systems. Proceedings of the National Academy of Sciences of the United States of America. 2017;**114**:9770-9778. DOI: 10.1073/pnas.1619807114

[26] Wigmore O, Mark B. Monitoring tropical debris-covered glacier dynamics from high-resolution unmanned aerial vehicle photogrammetry, cordillera

Blanca, Peru. The Cryosphere. 2017;**11**:2463-2480. DOI: 10.5194/tc-11-2463-2017

[27] Milner A. The Milner and Petts conceptual model of community structure within glacier-fed rivers: 20 years on. In: Research and Management for the 21st Century. River Science; 2016:156-170. DOI: 10.1002/9781118643525.ch8

[28] Füreder LN, G. Glacial stream ecology: Structural and functional assets. Water. 2020;**12**:376. DOI: 10.3390/w12020376

[29] Ilg C, Castella E. Pattern of macroinvertebrate traits along three glacial stream continuums. Freshwater Biology. 2006;**51**:840-853. DOI: 10.1111/j.1365-2427.2006.01533.x

[30] Lu Q, Zhang S, Du J, Liu Q, Dong C, Zhao J, et al. Multi-group biodiversity distributions and drivers of metacommunity organization along a glacial–fluvial–limnic pathway on the Tibetan plateau. Environmental Research. 2023;**220**(5):115236. DOI: 10.1016/j.envres.2023.115236

[31] Gobbi M, Lencioni V. Glacial biodiversity: Lessons from ground-dwelling and aquatic insects. In: Glaciers and the Polar Environment. London, UK: IntechOpen; 2020. DOI: 10.5772/intechopen.92826

[32] Lencioni V, Spitale D. Diversity and distribution of benthic and hyporheic fauna in different stream types on an alpine glacial floodplain. Hydrobiologia. 2015;**751**:73-87. DOI: 10.1007/s10750-014-2172-2

[33] Miserendino M, Brand C, Epele L, Di Prinzio C, Omad G, Archangelsky M, et al. Biotic diversity of benthic macroinvertebrates at contrasting glacier-fed systems in Patagonia Mountains: The role of environmental heterogeneity facing global warming. The Science of the Total Environment. 2017;**622-623**:152-163. DOI: 10.1016/j.scitotenv.2017.11.320

[34] Milner A, Brittain, J. Castella E, Petts G. Trends of macroinvertebrate community structure in glacier-fed rivers in relation to environmental conditions: A synthesis. Freshwater Biology. 2001; 46. 1833 - 1847. DOI: 10.1046/j.1365-2427.2001.00861.x.

[35] Milner A, Brown L, Hannah D. Hydroecological response of river systems to shrinking glaciers. Hydrological Processes. 2009;**23**:62-77. DOI: 10.1002/hyp.7197

[36] Espinosa R, Andino Guarderas P, Cauvy-Fraunié S, Dangles O, Jacobsen D, Crespo-Pérez V. Diversity patterns of aquatic macroinvertebrates in a tropical high-Andean catchment. Revista de Biología Tropical. 2020;**68**:S29-S53. DOI: 10.15517/rbt.v68iS2.44331

[37] Kuhn J, Andino Guarderas P, Calvez R, Espinosa R, Hamerlík L, Vie S, et al. Spatial variability in macroinvertebrate assemblages along and among neighbouring equatorial glacier-fed streams. Freshwater Biology. 2011;**56**:2226-2244. DOI: 10.1111/j.1365-2427.2011.02648.x

[38] Sudlow K, Tremblay S, Vinebrooke R. Glacial stream ecosystems and epilithic algal communities under a warming climate. Environmental Reviews. 2023;**31**:3. DOI: 10.1139/er-2022-0114

[39] Hotaling S, Finn D, Giersch J, Weisrock D, Jacobsen D. Climate change and alpine stream biology: Progress, challenges, and opportunities for the future. Biological reviews of the Cambridge Philosophical Society. 2017;**92**:2024-2025. DOI: 10.1111/brv.12319

[40] Molden D, Shrestha A, Immerzeel WW, Maharjan A, Rasul G, Wester P, et al. The great glacier and snow-dependent Rivers of Asia and climate change: Heading for troubled waters. In: Biswas AK, Tortajada C, editors. Water Security under Climate Change. Water Resources Development and Management. Singapore: Springer; 2022. DOI: 10.1007/978-981-16-5493-0_12

[41] Wayne R, Douglas M, Hecky R, Hershey A, Kling G, Lesack L, et al. Effects of climate change on freshwater of Arctic and subarctic North America. Hydrological Processes. 1997;**11**:873-902. DOI: 10.1002/(SICI)1099-1085(19970630)11:8<873::AID-HYP510>3.0.CO;2-6

[42] Young K, Lipton J. Adaptive governance and climate change in the tropical highlands of Western South America. Climatic Change. 2006;**78**:63-102. DOI: 10.1007/s10584-006-9091-9

[43] Bradley R, Vuille M, Diaz H, Vergara W. Threats to water supplies in the tropical Andes. Science (New York, N.Y.). 2006;**312**:1755-1756. DOI: 10.1126/science.1128087

[44] Durance I, Ormerod S. Climate change effects on upland stream macroinvertebrates over a 25-year period. Global Change Biology. 2007;**13**:942-957. DOI: 10.1111/j.1365-2486.2007.01340.x

[45] Bradley R, Keimig F, Diaz H, Hardy D. Recent changes in freezing level heights in the tropics with implications for the deglacierization of high mountain regions. Geophysical Research Letters. 2009;**36**:17. DOI: 10.1029/2009GL037712

[46] Carey M, Molden O, Rasmussen M, Jackson M, Nolin A, Mark B. Impacts of glacier recession and declining meltwater on mountain societies. Annals of the American Association of Geographers. 2016;**107**:1-10. DOI: 10.1080/24694452.2016.1243039

[47] Vuille M, Carey M, Huggel C, Buytaert W, Rabatel A, Jacobsen D, et al. Rapid decline of snow and ice in the tropical Andes – Impacts, uncertainties and challenges ahead. Earth-Science Reviews. 2017;**176**:195-213. DOI: 10.1016/j.earscirev.2017.09.019

[48] Mark BG, French A, Baraer M, Carey M, Bury J, Young KR, et al. Glacier loss and hydro-social risks in the Peruvian Andes. Global and Planetary Change. 2017;**159**:61-76. DOI: 10.1016/j.gloplacha.2017.10.003

[49] Kulkarni A, Mathur P, Rathore B, Alex S, Thakur N, Kumar M. Effect of global warming on snow ablation pattern in the Himalaya. Current Science. 2002;**83**:120-123

[50] Vergara W, Deeb A, Valencia A, Bradley R, Francou B, Zarzar A, et al. Economic impacts of rapid glacier retreat in the Andes. Eos, Transactions American Geophysical Union. 2007;**88**:261-264. DOI: 10.1029/2007EO250001

[51] Zambrano-Barragán C, Zevallos O, Villacis M, Enríquez D. Quito's climate change strategy: A response to climate change in the Metropolitan District of Quito, Ecuador. In: Otto-Zimmermann K, editor. Resilient Cities. Local Sustainability. Vol. 1. Dordrecht: Springer; 2011. DOI: 10.1007/978-94-007-0785-6_51

[52] Prasch M, Mauser W, Weber M. Quantifying present and future glacier melt-water contribution to runoff in a Central Himalayan river basin. The Cryosphere. 2013;**7**:889-904. DOI: 10.5194/tc-7-889-2013

[53] Rabatel A, Francou B, Soruco A, Gomez-Lopez J, Cáceres B, Ceballos J, et al. Current state of glaciers in the tropical Andes: A perspective on glacier evolution and climate change. Geophysical Research Abstracts. 2013;**15**:EGU2013-EGU8457

[54] Soruco A, Vincent C, Rabatel A, Francou B, Thibert E, Sicart J, et al. Contribution of glacier runoff to water resources of La Paz city, Bolivia (16° S). Annals of Glaciology. 2015;**56**:147-154. DOI: 10.3189/2015AoG70A001

[55] Nüsser M, Dame J, Sitara P, Kraus B, Baghel R, Schmidt S. Cryosphere-fed irrigation networks in the Northwestern Himalaya: Precarious livelihoods and adaptation strategies under the impact of climate change. Mountain Research and Development. 2019;**39**:R1-R11. DOI: 10.1659/MRD-JOURNAL-D-18-00072.1

[56] Schaner N, Voisin N, Nijssen B, Lettenmaier D. The contribution of glacier melt to streamflow. Environmental Research Letters. 2012;**7**:034029. DOI: 10.1088/1748-9326/7/3/034029

[57] Biemans H, Siderius C, Lutz A, Nepal S. Importance of snow and glacier meltwater for agriculture on the Indo-Gangetic plain. Nature Sustainability. 2019;**2**(7):594-601. DOI: 10.1038/s41893-019-0305-3

[58] Zulfiqar M, Abbasi I, Khan H, Nizami A, Hakeem A, Ali J, et al. Agricultural economy of Skardu is based on glaciers and snow melting – A case study of Burgay watershed. Sarhad Journal of Agriculture. 2019;**35**(2):336-341. DOI: 10.17582/journal.sja/2019/35.2.336.341

[59] Ashraf A, Batool A. Evaluation of glacial resource potential for sustaining kuhl irrigation system under changing climate in the Himalayan region. Journal of Mountain Science. 2019;**16**:1150-1159. DOI: 10.1007/s11629-018-5077-0

[60] Chevallier P, Pouyaud B, Suarez W, Condom T. Climate change threats to environment in the tropical Andes: Glaciers and water resources. Regional Environmental Change. 2010;**11**:179-187. DOI: 10.1007/s10113-010-0177-6

[61] Buytaert W, Moulds S, Acosta L, De Bièvre B, Olmos C, Villacis M, et al. Glacier melt content of water use in the tropical Andes. Environmental Research Letters. 2017;**12**:114014. DOI: 10.1088/1748-9326/aa926c

[62] Cauvy-Fraunié S, Condom T, Rabatel A, Villacis M, Jacobsen D, Dangles O. Technical note: Glacial influence in tropical mountain hydrosystems evidenced by the diurnal cycle in water levels hydrology and earth system sciences. Hydrology and Earth System Sciences. 2013;**17**:4803-4816. DOI: 10.5194/hess-17-4803-2013